AF302105

Melanie Syha

Microstructure evolution in strontium titanate

Investigated by means of grain growth simulations
and x-ray diffraction contrast tomography experiments

**Schriftenreihe
des Instituts für Angewandte Materialien
*Band 40***

Karlsruher Institut für Technologie (KIT)
Institut für Angewandte Materialien (IAM)

Eine Übersicht über alle bisher in dieser Schriftenreihe erschienenen Bände
finden Sie am Ende des Buches.

Microstructure evolution in strontium titanate

Investigated by means of grain growth simulations and x-ray diffraction contrast tomography experiments

by
Melanie Syha

Dissertation, Karlsruher Institut für Technologie (KIT)
Fakultät für Maschinenbau
Tag der mündlichen Prüfung: 10.04.2014

Impressum

Scientific
Publishing

Karlsruher Institut für Technologie (KIT)
KIT Scientific Publishing
Straße am Forum 2
D-76131 Karlsruhe

KIT Scientific Publishing is a registered trademark of Karlsruhe
Institute of Technology. Reprint using the book cover is not allowed.

www.ksp.kit.edu

Print on Demand 2014

ISSN 2192-9963
ISBN 978-3-7315-0242-5
DOI   10.5445/KSP/1000042175

Danksagung

Die vorliegende Dissertation entstand während meiner Tätigkeit als wissenschaftliche Mitarbeiterin am Institut für Angewandte Materialien des Karlsruher Instituts für Technologie.

Mein besonderer Dank gilt Herrn Prof. Dr. Peter Gumbsch für die sehr gute Betreuung, die Übernahme des Hauptreferats und das große fachliche Interesse an dieser Arbeit. Herrn Prof. Dr. Michael Hoffmann möchte ich für die Übernahme des Korreferats danken.

Herrn Dr. Daniel Weygand danke ich für die gute fachliche Betreuung, sowie zahlreiche wissenschaftliche Diskussionen. Dr. Wolfgang Rheinheimer und Dr. Michael Bäurer vom Teilinstitut Keramik im Maschinenbau danke ich für die gute Zusammenarbeit bei der Probenpräparation.

Barbara Lödermann, Andreas Trenkle und Werner Augustin haben durch ihre professionelle Arbeitsweise, ihre außerordentliche Teamfähigkeit und ihren feinen Humor nachhaltig zum Gelingen dieser Arbeit beigetragen. Ihnen gebührt besonderer Dank. Dr. Rudolf Baumbusch danke ich für viele inspirierende Gespräche. Seine geradlinige Art und der Mut auch unangenehme Dinge auszusprechen beeindrucken mich. Stellvertretend für alle wissenschaftlichen Hilfskräfte bedanke ich mich bei Philipp Bender.

Mein herzlicher Dank gilt darüberhinaus Dr. Wolfgang Ludwig von der European Synchrotron Radiation Facility (ESRF), Grenoble sowie Dr. Erik Lauridsen vom Risø National Laboratory for Sustainable Energy, Roskilde für die Unterstützung bei den Röntgenbeugungstomographiemessungen und die angenehme Arbeitsatmosphäre. Desweiteren möchte ich dem Karlsruher House of Young Scientists (KHYS) für die finanzielle Unterstützung der Auslandsaufenthalte in Frankreich und Dänemark danken.

Dr. Andreas Graff sowie Katrin Reinhardt vom Fraunhofer Institut für Werkstoffmechanik in Halle danke ich für die Unterstützung bei den Elektronenrückstreubeugungsmessungen.

Ganz besonderer Dank gilt Christoph Dahnz, Jasmin Wild, sowie meinen Eltern, ohne deren Geduld und Unterstützug diese Arbeit nicht möglich gewesen wäre.

*"By the pricking of my thumbs,
Something wicked this way comes."*
Macbeth Act 4

scene 1, 44-49

Abstract

Understanding the physical processes which lead to microstructure formation during fabrication and annealing of ceramic materials is a long sought goal among material scientists. Using strontium titanate as a model system for perovskite ceramics, the present work combines advanced non-destructive 3D characterization techniques and computational modeling of microstructure evolution in order to link grain morphology, interface anisotropy and microstructure evolution to macroscopic physical properties.

The effects of orientation dependent interface properties on growth kinetics are studied both experimentally and in parameter variations using a front tracking model. The microstructure of a polycrystalline strontium titanate specimen is characterized at two stages during microstructure evolution by means of X-ray Diffraction Contrast Tomography. In between the tomography scans, the specimen is annealed for 1h at 1600°C. The microstructure reconstruction of the specimen in the annealed state is validated against electron backscatter diffraction investigations.

The individual and combined effects of anisotropy of both grain boundary energy and mobility are investigated in simulations using a three dimensional vertex dynamics model with systematically varied parameters. Both growth kinetics and microstructure are found to be strongly influenced by the orientation dependent grain boundary energy, while only a weak impact of the grain boundary mobility was observed even for strong grain boundary mobility anisotropy. A measurable effect on the microstructure is only detected, when mobility anisotropy is combined with grain boundary energy effects. Then abnormal grain growth can be observed. A mechanism for the nucleation of abnormal growth in candidate grains is proposed.

Kurzzusammenfassung

Die makroskopischen Eigenschaften polykristalliner keramischer Werkstoffe werden maßgeblich durch die Mikrostrukturentwicklung beeinflusst. Das bisherige Wissen über die während der Mikrostrukturentwicklung ablaufenden Prozesse, insbesondere die Verknüpfung mikrostruktureller und makroskopischer Materialparameter ist nach wie vor unvollständig. Die vorliegende Arbeit kombiniert zerstörungsfreie 3D Charakterisierungsmethoden und rechnergestützte Modellierung der Mikrostrukturentwicklung. Dabei wird Strontiumtitanat als Modellsystem für Perowskitkeramiken betrachtet. Ziel ist, die Zusammenhänge zwischen Morphologie, anisotropen Grenzflächeneigenschaften und Mikrostrukturentwicklung zu verstehen und so die Auswirkungen auf makroskopische Werkstoffkennwerte beschreiben zu können.

Der Einfluss orientierungsabhängiger Grenzflächeneigenschaften auf die Wachstumskinetik wird sowohl experimentell, als auch mittels systematischer Parametervariation eines Grenzflächenmodells untersucht. Die Mikrostruktur einer polykristallinen Strontiumtitanat-Probe wird an zwei definierten Zeitpunkten innerhalb eines Auslagerungsprozesses mittels Röntgenbeugungs-Tomographie in 3D charakterisiert. Zwischen den beiden Aufnahmen findet eine Auslagerung für eine Stunde bei 1600°C statt. Für die Mikrostruktur-Rekonstruktion im augelagerten Zustand erfolgt eine Validierung mittels Elektronen-Rückstreu-Diffraktion.

Parallel dazu werden Kornwachstumssimulationen unter systematischer Variation der Grenzflächenenergie sowie -mobilität durchgeführt. Während die Variation der Grenzflächenenergie einen großen Einfluss auf die Mikrostruktur und die Wachstumsdynamik ausübt, findet sich nur ein schwach ausgepägter Zusammenhang zwischen Mikrostrukturentwicklung und Anisotropie der Grenzflächenmobilität. Diese hat nur in Kombination mit Grenzflächenenergieeffekten messbare Auswirkungen auf die Struktur. Basierend auf diesen Ergebnissen wird ein Nukleationsmechanismus für abnormal wachsende Körner innerhalb einer isotropen Matrix vorgeschlagen.

Symbols

(ϕ, θ, ψ)	Euler angles
$\mathbf{n} = (n_1, n_2, n_3)$	normal vector
$\mathbf{t} = (t_1, t_2, t_3)$	translation vector of neighbouring crystallites
Φ	misorientation of neighbouring crystallites
E_D	elastic energy of edge dislocation per unit length
μ	shear modulus
ν	Poissons ratio
r_0	radius of the dislocation core
$\mathbf{b}$	Burgers vector
b	norm of the Burgers vector
d	dislocation spacing
E_C	energy of the dislocation core
$\gamma(\Phi)$	misorientation dependent grain boundary energy per unit length
θ	inclination angle
$\sigma(\theta)$	orientation dependent surface energy
$\sigma(\mathbf{n})$	normal dependent surface energy
M	grain boundary mobility
$m(\mathbf{n})$	surface normal orientation dependent mobility
$\mathbf{v}$	boundary velocity
$\mathbf{P}$	driving force for grain boundary motion
M_L	low angle grain boundary mobility
M_H	high angle grain boundary mobility
T	temperature
D_L	lattice self-diffusion coefficient
Ω	atomic volume
k_B	Boltzmann constant
Q	activation energy
R	universal gas constant
N_P	number of polyhedra in observed volume
N_F	number of faces

N_E	number of edges
N_V	number of (quadruple) vertices
γ_i	interface tension
$\mathbf{b}_i$	in-plane direction of the interface
t	time
P	polyhedron
$L(P)$	mean width of polyhedron
$\sum_{i=1}^{n} e_i$	total edge length of the polyhedron
β_i	exterior turning angle
G	total free energy
R	mean grain radius
α	proportionality factor
	mean symbol
β	local coefficient
λ	metrical scale factor
d_{hkl}	spacing of hkl-planes
Θ	angle between incident ray and scattering plane
λ	X-ray wavelength
n	reflection order
hkl	lattice plane type
$\Delta\sigma$	cusp depth
ϕ	opening angle of cusp
V	total grain boundary energy
$\mathbf{r}$	vector of interface position
$\mathbf{v}$	vector of interface velocity
R	energy dissipation
L	sidelength of simulated cube
l_{crit}	critical reference length
n_{virt}	number of discretizational vertices along triple line
$\mathbf{D}^A, \mathbf{D}^{ABC}$	Lagrange coupling matrices
f_j^{T}	torque contribution to nodal force
$\mathbf{R}$	rotation matrix
$\mathbf{f}^A$	nodal force of node A

$\mathbf{v}^A$	nodal velocity of node A
Δt	time step, small time interval
κ	friction factor for virtual vertices on flat boundary
l	subedge length
Δl_{max}	change of subedge length
$M(\mathbf{n})$	grain boundary mobility potential
$m(\mathbf{n})_l, m(\mathbf{n})_r$	normal dependent surface mobility
χ^2	probability distribution
E_i, O_i	expected and observed frequencies for event i
σ_S	standard deviation
μ_S	mean value
$\overline{F}_G$	average number of neighbors of a grain
$\overline{E}_G$	average number of edges per grain
Ψ	sphericity
$A(t)$	area
V_G	grain volume
O_G	grain surface area
k	normalized growth rate
K	relative growth rate
A_{tot}	total surface area of the pole sphere
A_C	pole sphere surface area affected by energy cusp
f_C	fraction of grain faces affected by energy cusp
f_P	fraction of grain faces affected by mobility peak
f_{GF}	fraction of high mobility grain faces
g_C	average characteristic grain size
$\mathbf{f}^G_{det}$	detachment force
$\mathbf{f}^G_{pin}$	opposing force imposed by new grain boundaries

Contents

1. Introduction

Tailoring material properties to specific application requirements is one of the major challenges in materials science. Since most physical and mechanical properties of polycrystals are controlled by size and composition of the microstructure, it is crucial to understand the processes leading to microstructure formation. Grain boundary motion, forming the morphology and topology of the grain boundary ensemble, is one of the dominant processes during microstructure evolution. Hence a thorough comprehension of grain growth and its underlying mechanisms is fundamental.

Due to their mechanical, thermal and chemical stability as well as their unique electrical, optical and magnetic properties, ceramics are one of the technologically most relevant material classes. Ceramic materials with perovskite-type structure can easily be modified by appropriate changes in composition. That way a large number of chemical compounds that are of considerable technological importance can be produced. Metal oxides of perovskite structure such as $BaTiO_3$, $PbTiO_3$ or $Pb(ZrTi)O_3$ are used as dielectric for manufacturing multi-layer capacitors, thermistors, transducers, temperature sensors, piezoelectric transducers and high temperature superconductors.

Since the materials properties are strongly influenced by the processing route, the major aim of this thesis is to contribute to the long term goal of quantitative investigation of microstructure formation in perovskite ceramics. Here, the correlation between grain boundary properties and macroscopic materials properties is of particular interest. Throughout this thesis, polycrystalline bulk strontium titanate ($SrTiO_3$) serves as a model material for functional ceramics with a perovskite (ABO_3) structure. Strontium titanate $SrTiO_3$ was chosen as model system due to the absence of phase transitions in the temperature regime of interest and the rather extensive amount of experimental work focussing on the anisotropy of interface properties [1–5] and growth kinetics [6,7] in this material. Apart from atomistic modeling of specific boundaries [8] there have been no modeling approaches to assess this topic. Moreover, a recently discovered deviation from Arrhenius type growth behavior in strontium titanate [7] is still unexplained.

The investigation of the complex processes occuring during mirostructure evolution in any polycrystalline material poses several challenges to material scientists: Annealing experiments giving insight into the real microstructure are time, labor and cost intensive. The restriction to destructive two dimensional (2D) characterization methods rules out an observation of the same structure at various stages during microstructure evolution. And even if non-destructive 3D characterization is feasible, high temperatures or long annealing times prohibit a continuous observation of the evolving structure. Growth simulations on the other hand are adaptable in time and space resolution of the characterization but they rely to a large extent on simplification and generalization of both structural properties and migration mechanisms. Moreover they are dependent on input parameters generated from experiments or more fundamental simulations.

One major aim of this thesis is to integrate experimental and modeling efforts in the characterization of microstructure evolution in strontium titanate. By generating dedicated input data for the model approach of choice and using the models efficiency for parameter variations, a better understanding of the link between grain morphology, microstructure evolution and macroscopic materials properties shall be established. Therefore, experimental and modeling investigations are combined in the present work: Grain growth simulations are used to study the influence of individual interface properties that are not easily controlled experimentally. Eventually they allow to predict the properties for materials manufactured under distinct conditions. Annealing experiments using three dimensional x-ray imaging provide realistic microstructures and valuable insight in orientation dependent grain boundary properties, serving both as validation and indispensable source for input parameters for grain growth simulations.

2. Literature

This chapter gives an overview of the most relevant aspects required to discuss grain boundary motion and its modeling in perovskite ceramics. A brief introduction into grain boundaries and their properties is followed by a review of the fundamental principles of grain boundary motion with special regard to ceramic materials. Subsequent subchapters cover computational growth models and a short overview over microstructure characterization techniques.

2.1. Grain Boundaries

A grain boundary is defined as the interfacial transition region between two crystals of the same phase and crystallographic structure but different orientation with regard to a global coordinate system. Their nature and behavior influences microstructure formation and thus the macroscopic physical and mechanical properties of any polycrystalline material substantially.

2.1.1. Grain Boundary Properties

The overall geometry of a grain boundary requires eight parameters for a full mathematical description in three dimensions: three Euler angles (ϕ, θ, ψ) to describe the orientation relationship between the two adjacent crystallites, two components of the normal vector $\mathbf{n} = (n_1, n_2, n_3)$ defining the spatial orientation of the interface with respect to one of the two neighboring grains and three components of the translation vector $\mathbf{t} = (t_1, t_2, t_3)$ that describes the displacement of the two crystallites. Equivalent to using the three Euler angles describing the rotation of one lattice into the other is a description giving an axis and angle of rotation. The minimal rotational angle Φ (using the crystal symmetry) needed to rotate one set of crystal axes into coincidence with the other crystal is called misorientation. Interfaces with a small misorientation ($\Phi < 15°$) are considered low angle grain boundaries. Interfaces between crystallites with a larger misorientation are referred to as high angle grain boundaries.

Each grain boundary is associated with specific kinetic and thermodynamic grain boundary properties like mobility, energy and entropy, which depend on crystallography and are in principle a function of the eight geometry parameters [9]. Thereof, the microscopic parameters $\mathbf{t} = (t_1, t_2, t_3)$ that describe the relative displacement of the two crystallites are assumed to adjust themselves such that the total energy reaches a local minimum. Accordingly, the number of controllable degrees of freedom of the boundary reduces to five. The knowledge of these grain boundary properties is a crucial prerequisite for the simulation of microstructure evolution.

Grain Boundary Energy

Low angle grain boundaries can be regarded as formed of dislocations. Their interface free energy can be calculated following Read and Shockley [10] as the energy of an array of edge dislocations. The energy of an edge dislocation per unit length is given by

$$E_D = \frac{\mu b^2}{4\pi(1-\nu)}\ln\frac{d}{r_0} + E_C, \qquad (2.1)$$

where μ and ν are the shear modulus and Poissons ratio, $r_0 \approx b$ is the radius of the dislocation core (b is the norm of the Burgers vector $\mathbf{b}$), d is the dislocation spacing. The elastic contribution is given by the first term in equation 2.1 while the second term E_C gives the energy of the dislocation core.

When the boundary is considered as an array of parallel edge dislocations, the dislocation spacing d defines the angular rotation Φ

$$sin(\Phi/2) = \frac{b}{2d}, \qquad (2.2)$$

which for small angles is

$$\Phi \approx \frac{b}{d}. \tag{2.3}$$

Accordingly, the misorientation dependent grain boundary energy per unit area can be expressed as [11]

$$\gamma(\Phi) = \frac{\Phi}{b} \left[\frac{\mu b^2}{4\pi(1-\nu)} \ln\frac{1}{\Phi} + E_C \right]. \tag{2.4}$$

For high angle grain boundaries, the dislocation model fails to predict the grain boundary energy correctly, because there is no unique dislocation arrangement and the dislocation cores tend to overlap beyond a rotation angle of $10 - 15°$. The energy of high angle grain boundaries is found to be higher than that of low angle boundaries and approximately constant except for some particular orientations for which the atomic fit is better than in general and the corresponding grain boundary energy is low. Typically, these special boundaries are described in terms of the so called coincidence site lattice (CSL) concept [12].

To more precisely assess the energy of high angle grain boundaries atomic scale computer simulations can be employed for distinct oriented interfaces. Other methods, like capillarity vector reconstruction [13] or inverse Wulff construction [14] from pore shapes aim at identifying orientation dependent relative surface energies and make the transition to interface energies by assuming that anisotropic surface energies and interface energies are closely related. Interface energies are then assumed proportional to surface energies reduced by an interface binding term [15, 16].

Often, it is sufficient, to describe the energy γ of a grain boundary with fixed crystal misorientation as a function of its inclination angle θ, as it is done in the Wulff plot [14], see figure 2.1. Here, a polar plot of the orientation dependent surface energy $\sigma(\theta)$ of the crystal for each inclination is represented by a vector in the direction of the surface normal. The Wulff plot may show distinct energy minima, so called cusps, at energetically favorable inclinations. Using energy minimization arguments, the equilibrium shape of a crystal can then be determined by the Wulff construction [14]. Therefor planes perpendicular to each $\sigma(\theta)$ vector are drawn at the point, where the vector intersects the Wulff plot. Then the

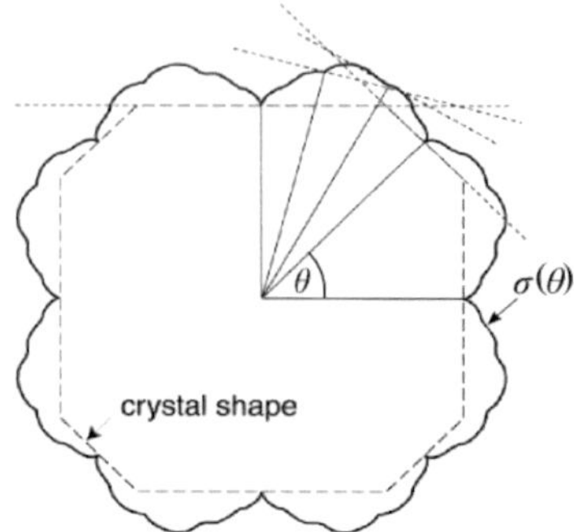

Figure 2.1.: Cross section of a 3D Wulff plot showing the energy σ of a surface with fixed crystal misorientation as a function of its inclination angle θ [17].

shape of the crystal in the equilibrium state is represented as the inner envelope of these planes.

Grain Boundary Mobility

Grain boundary mobility M is commonly defined as the proportionality constant between the boundary velocity $\mathbf{v}$ and the driving force for grain boundary motion $\mathbf{P}$:

$$\mathbf{v} = M\mathbf{P}. \qquad (2.5)$$

Here, the driving force $\mathbf{P}$ can be of various nature e.g. stored energy differences, plastic deformation or temperature gradients. Since grain boundary motion is a thermally activated process, M generally exhibits an Arrhenius-like temperature dependency. It has been shown in experiments, that the grain boundary mobility is also a function of composition and crystallographic type of the material [18]. Pronounced anisotropy with respect to misorientation and boundary plane orientation was observed for the grain boundary mobility in metals [9, 19].

On a basic level and to keep the analogy to the description of grain boundary energy, it is possible to distinguish between the mobility of low angle and high angle grain boundaries: Since low angle grain boundaries migrate by a combination of dislocation glide and dislocation climb, their mobility M_L is a function of the grain boundary structure. Assuming, that the rate of dislocation climb is diffusion limited [11] the mobility of low angle grain boundaries can be approximated as [20]

$$M_L \approx \frac{8D_L\Omega d}{k_B T c b^2},$$ (2.6)

where d denotes the spacing between the dislocations, T the temperature and c is the correlation factor for atomic jumps in diffusion. D_L is the self-diffusivity which follows an Arrhenius temperature dependence and Ω is the atomic volume associated with the grain boundary when a constant boundary thickness is assumed. The mobility of high angle grain boundaries M_H is based on thermally activated atomic transfer across the interface and is often found to obey an Arrhenius type relationship of the form

$$M_H = M_0 \exp\left(-\frac{Q}{RT}\right),$$ (2.7)

where Q is the activation energy and R the universal gas constant.

Topology

In addition to the structure and properties of individual grain boundaries some attention must be drawn to their arrangement inside the material. Influencing factors on the formation of a metastable configuration are the topological requirement of a space filling grain boundary network on one side and boundary tension balance on the other side.

In three dimensions, space filling polyhedra fulfill the Euler relation

$$N_V - N_E + N_F - N_P = 1,$$ (2.8)

where N_P is the number of polyhedra contained in the observed volume, N_F is the total number of faces, N_E the total number of edges and N_V the number of (quadruple) vertices.

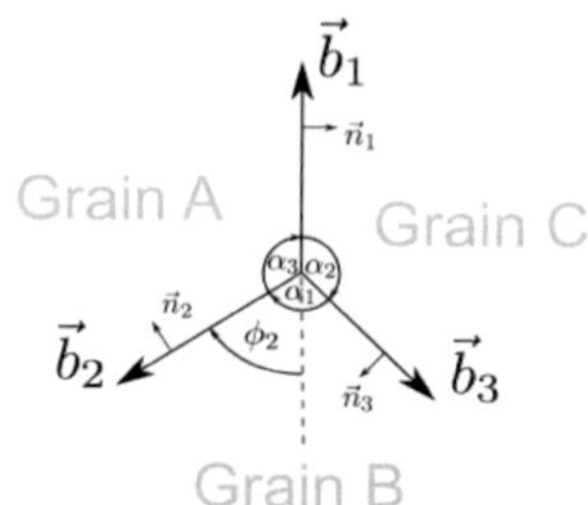

Figure 2.2.: Equilibrium of boundary tensions at a triple line.

Moreover, the grain boundaries seek to arrange in a way such that the equilibrium equation for boundary tension, the so called Herring relation [21]

$$\sum_{i=1}^{3} \left[\gamma_i \mathbf{b}_i + \left(\frac{\partial \gamma_i}{\partial \Phi_i} \right) \mathbf{n}_i \right] = \mathbf{0}, \qquad (2.9)$$

is fulfilled. Here, γ_i marks the boundary tension, $\mathbf{b}_i$ the in-plane direction of the boundary and $\mathbf{n}_i$ its normal direction, while the inclination of the boundary with respect to the crystal coordinate system is denoted with Φ_i, see figure 2.2. There is however no formation of three dimensional (3D) plane-faced polyhedra that fulfills equation 2.9 and is space filling at the same time [22]. In order to fulfill these two requirements simultaneously grain boundaries must be curved.

Real polycrystalline microstructures consist of grains of different shapes and sizes. Still, for materials with uniform grain boundary energy, the dihedral angles are found to be close to 120°.

For materials with anisotropic grain boundary energy a torque contribution resulting from the second part of the Herring relation 2.9, increases the dihedral angle opposite to the boundaries with comparatively lower energy.

2.1.2. Grain Boundary Motion

Grain boundary motion can be described as the motion of a grain boundary relative to the crystal lattices sharing the boundary. In the absence of long range diffusional fluxes, grain boundary motion can be imagined as one crystallite growing at the expense of another one. Most often, this is described as atoms being transferred locally across the interface [11] or a generation of lattice sites at the surface of the growing grain and conversely a destruction of lattice sites at the surface of the shrinking grain [9]. Effectively, grain boundary motion comprises the non-zero net exchange of lattice sites across the boundary.

While there is no unified theory of grain boundary migration in literature, it is generally agreed upon that motion occurs as response to a driving force

$$\mathbf{P} = - \triangledown G \tag{2.10}$$

that exists, whenever the grain boundary motion causes the system total free energy G to decrease.

There are a number of sources leading to driving forces for grain boundary motion. These range from applied stress, over bulk free energy differences during phase transformations in heterophase boundaries to average stored energy of plastic deformation during recrystallization. The most relevant driving force for boundary motion however is the reduction of the total grain boundary area. This driving force controls temperature activated grain growth processes like annealing and is the only driving force regarded in this work.

2.2. Growth Kinetics of Polycrystals

Grain growth in polycrystals can be regarded as the collective behavior of the grain boundaries during isothermal annealing. Typically, two types of growth behavior are distinguished: "normal grain growth" during which the normalized grain size distribution remains uniform, and "abnormal grain growth" resulting in a bimodal grain size distribution.

2.2.1. Normal Growth

The curvature driven growth kinetics of a system of crystallites during normal growth is well described by the simple kinetic equation

$$\frac{dA}{dt} = \alpha \gamma M \qquad (2.11)$$

proposed by Hillert [23]. Integrating this differential equation gives the parabolic growth law

$$A(t) = A(0) + \alpha \gamma M t, \qquad (2.12)$$

where $A(t)$ is the mean grain size evaluated as cross section at any given time t and $A(0)$ is the mean grain size for $t = 0$, α is a geometrical factor taking into account the grain boundary network. This factor is found to be of the order of unity [23, 24]. During normal grain growth grain structures are of uniform appearance and the normalized grain size distribution remains uniform. Often, a linear growth behavior can be assumed [25]. Then, the normalized area evolution

$$\frac{A(t)}{A(0)} = 1 + \frac{\alpha \gamma M}{A(0)} t \qquad (2.13)$$

allows to extract a linear growth rate k

$$k = \frac{\alpha \gamma M}{A(0)}. \qquad (2.14)$$

Two alternate expressions for grain size distributions have been proposed based on grain growth theory. Hillert derived an expression for the size distribution from the mean field model [23]:

$$f_H(u) = C \frac{u}{(2 - u)^{2+\delta}} \exp\left(\frac{2\delta}{(2 - u)}\right). \qquad (2.15)$$

Here, C is a constant, δ is a dimensionality factor being 2 for 2D growth and 3 for 3D grain growth and u is defined as $u = \frac{R}{R_{cr}}$ with at cut off at $u = 2$. R is the grain radius and R_{cr} is the critical grain radius needed for growth. Louat on the other hand assumed that grain growth can be regarded as random diffusion-like process and expressed the grain size distribution function [26] as

$$f_L(R) = \frac{\pi}{2} R \exp\left(-\frac{\pi}{4} R^2\right). \tag{2.16}$$

Feltham used a log-normal function to describe experimentally observed grain size distributions in metals [27]:

$$f_{ln}(R) = \frac{Z}{\sqrt{\pi} R} \exp\left(-Z^2 \left(\ln\left(\frac{R}{\overline{R}}\right)\right)^2\right). \tag{2.17}$$

Here, Z denotes a constant and $\overline{R}$ is the median grain radius. Despite the fact that there is no physical reasoning for the log-normal distribution, this function is widely used to describe experimentally observed grain size distributions of dense single phase polycrystals [28–30].

Since the curvature needed to fulfill the Herring relation and the space filling restrictions of a realistic material contribute to the driving force for migration of grain boundaries, grain growth can be considered a topological process. In 2D and for isotropic conditions, the von Neumann-Mullins relation [31, 32] gives the area change rate of an s-sided grain as

$$\frac{dA}{dt} = \frac{\pi}{3} M\gamma(s - 6). \tag{2.18}$$

More recently, this relation has been generalized to three dimensions [33]. MacPherson and Srolovitz showed that the volume change rate of any given discretized polyhedron P can be expressed as a function of its topology [33]

$$\frac{dV}{dt} = -2M\pi\gamma \left(L(P) - \frac{1}{6} \sum_{i=1}^{n} e_i(P)\right), \tag{2.19}$$

where $\sum_{i=1}^{n} e_i(P)$ is the total edge length of the polyhedron. $L(P)$ is denoted as mean width. For a triangulated polyhedron it is given by

$$L(P) = \frac{1}{2\pi} \sum_{i=1}^{m} \epsilon_i \beta_i \tag{2.20}$$

where ϵ_i is the length of an edge shared by two triangles and β_i is the exterior turning angle these triangles enclose. The mean width can be interpreted as mean curvature of the polyhedron.

2.2.2. Abnormal Growth

During abnormal grain growth a small number of grains grow at an above-average growth rate at the expense of the remaining (so called matrix) grains. The resulting bimodal grain size distribution is only eliminated when, after a certain time, all the matrix grains have been consumed.

Conditions for the onset of abnormal grain growth have been discussed for decades. While earlier works [23, 34] assuming uniform grain boundary energies focus on the stability of cellular structures in order to explain the nucleation of abnormal growth, more recent publications consider abnormal grain growth as being attributed to anisotropies in grain boundary properties [35, 36]. Both analytical and simulation studies following this approach typically consider idealized cellular microstructures comprised of two types of grains: matrix grains with mean properties (radius $\overline{R}$, grain boundary mobility and energy $\overline{M}$, $\overline{\gamma}$ and misorientation $\overline{\theta}$) and candidate grains for abnormal growth, which may deviate significantly from the matrix grains in one or more of these properties. Assuming that no grain rotation or grain sliding takes place, the driving force for the growth of such a candidate grain in an otherwise homogeneous matrix can be derived as

$$P = \frac{1.5\beta\overline{\gamma}}{\overline{R}} - \frac{2\alpha\gamma}{R}, \tag{2.21}$$

assuming a reduction in energy through growth. Since the net driving force for the special case $\overline{R} = R$ is zero, the relation between the constants α and β is found to be $\beta = \frac{4}{3}\alpha$. Equation 2.21 can be rewritten to

$$P = 2\alpha \left(\frac{\overline{\gamma}}{\overline{R}} - \frac{\gamma}{R} \right). \tag{2.22}$$

Typically, α is of the order of unity. Keeping this in mind and following the generally accepted approach for the calculation of grain boundary motion, the growth rate $\frac{dR}{dt}$ for the candidate grain is written as

$$\frac{dR}{dt} = MP = M \left(\frac{\overline{\gamma}}{\overline{R}} - \frac{\gamma}{R} \right) \tag{2.23}$$

using the driving force as derived in equation 2.21. Likewise, the growth rate of any given matrix grain is given as [23]

$$\frac{d\overline{R}}{dt} = \frac{\overline{M\gamma}}{4\overline{R}}.$$

(2.24)

Thompson *et al.* [34] suggested that the condition for abnormal growth of the candidate grain can be written as

$$\frac{d\left(\frac{R}{\overline{R}}\right)}{dt} > 0.$$

(2.25)

With

$$\frac{d\left(\frac{R}{\overline{R}}\right)}{dt} = \frac{1}{\overline{R}^2}\left(\overline{R}\frac{dR}{dt} - R\frac{d\overline{R}}{dt}\right) > 0$$

(2.26)

and equation 2.23 for the growth rate $\frac{dR}{dt}$ of the candidate grain and 2.24 for the growth rate $\frac{d\overline{R}}{dt}$ of a matrix grain, equation 2.25 can be rewritten to

$$M\gamma - \frac{\overline{R}M\gamma}{R} - \frac{R\overline{M\gamma}}{4\overline{R}} > 0.$$

(2.27)

Therefore, according to these approaches, the nucleation of abnormal growth is dependent on relative grain size as well as relative grain boundary energy and mobility of the candidate grains.

Topological approaches to abnormal grain growth [37–40] explicitly examine the number of faces or the boundary curvature of candidate grains in order to identify a topological analogon to inequality 2.27. Dealing with 3D structures composed of irregular polyhedra however makes this task challenging and necessitates several simplifying assumptions. The most recent results [39] present a criterion for abnormal grain growth solely depending on the number of faces N_F a candidate grain contains. This work is based on the method of representing irregular 3D networks using infinite sets of regular polyhedra [41, 42]. For these regular polyhedra, it was shown, that their volume V is proportional to the number of faces N_F they contain

$$V^{\frac{1}{3}} \propto N_F^{\frac{1}{2}}\lambda,$$

(2.28)

with λ as a metrical scale factor. Then, the obvious deduction, that the basic criterion for abnormal growth stated in equation 2.25 can be expressed in terms of grain volumes can be extended to

$$\frac{d\left(R/\overline{R}\right)}{dt} = \frac{d\left(V/\overline{V}\right)^{\frac{1}{3}}}{dt} = \frac{d\left(N_F/\overline{N_F}\right)^{\frac{1}{2}}}{dt} > 0. \tag{2.29}$$

2.3. Computational Growth Models

Aiming at the quantitative investigation and prediction of microstructure evolution, grain growth models are a powerful tool for the investigation of properties that are not easily controlled experimentally. Typically operating on the mesoscopic scale, the classical approaches for the simulation of grain growth can be categorized as either stochastic or deterministic. The most commonly applied approaches are Monte Carlo (stochastic), Phase Field and Vertex Dynamics (both deterministic) models. All models apply a discretization of the grain structure on a planar or volumetric basis, treating the grain boundaries as dividing interfaces between regions of different orientations.

Monte Carlo

In Monte Carlo models [43–47] the microstructure is represented as equally sized voxels associated with orientation information. Accordingly, grain boundaries occur between pairs of voxels having different orientations and grain boundary properties can be assigned as a function of misorientation. The time evolution of these models is driven by energy minimization and the evolution is performed voxel-wise. Starting from a randomly selected voxel, the change in the systems total energy induced by the change of state of the particular voxel is calculated and accepted according to a Metropolis Scheme [48].

Phase Field

Phase Field models for the simulation of grain growth [49–52] typically apply a regular volumetric discretization. In these models, crystallographic orientations are expressed through order parameters. In contrast to the sharp interface models, these models show a continuous transition of the order parameters. Groups of voxels of similar orientation make up grains, a finite interface width is achieved by a gradient penalty for the order parameters of a grain. The microstructure evolution is controlled through the variation of a free-energy potential with respect to the order parameters. Interface normal dependent grain boundary properties can be implemented [52].

Vertex Dynamics

Vertex Dynamics models [24, 53] are sharp interface models. Discretizing the interfaces only allows to attribute grain boundary properties as a function of all five degrees of freedom to every basic discretizational unit. Typically the grain boundary mesh is represented by a triangulation. A detailed description of the Vertex Dynamics Model applied in the present work is provided in section 4.1.

All above mentioned models have been applied to the simulation of grain growth in ideal systems. Due to their different approaches each of the models is suited to serve different purposes:
Vertex models show good agreement with theory in terms of scaling and growth laws. Therefore they are applied to simulate ideal normal grain growth [24, 54, 55] and triple line/quadruple point drag [56, 57]. More recently, systems with inclination and misorientation dependent grain boundary properties [58] and the nucleation of abnormal grain growth [59] have been modeled as well. Phase field models are computationally demanding but the only class of models that allow to simulate phenomena in which diffusion or long-range stresses are coupled to the grain growth process. Examples include simulations studying the effect of porosity on grain growth kinetics in polycrystalline ceramics [60], plastic flow [61] or creep deformation phenomena [62]. Monte Carlo models have been used

for large systems with complex textures and second phase particles such as Zener pinning [63]. This type of model has also been applied to the simulation of recrystallization during annealing of deformed metals [64]. Still, the applicability of all models to real materials is limited by the lack of experimental data i.e. input structures as well as grain boundary properties.

2.4. Microstructure Characterization

The characterization and quantitative analysis of microstructures in polycrystals is of great importance to materials scientists, since structure defining quantities such as grain size or homogeneity have a strong impact on the physical properties of the material.

For decades, microstructure characterization was restricted to destructive two dimensional (2D) metallography applying optical microscopy, electron microscopy or X-ray microscopy to manually polished cross-sections of the material. In the early 1990ies, albeit tedious, serial sectioning [65] allowed first steps towards a 3D microstructure characterization. Today roboted serial sectioning [66], focused ion beam (FIB) milling in dual beam instruments [67, 68] and femtosecond-laser based ablation techniques [69] allow for destructive access to full 3D crystallographic information. Additionally, non-destructive, 3D X-ray characterization techniques were proposed at the beginning of the current century. Among these, 3D X-ray diffraction microscopy (3DXRD) [70–72] and X-ray diffraction contrast tomography (DCT) [73, 74] as well as differential aperture X-ray microscopy [75] allow for non-destructive acquisition and reconstruction of 3D grain microstructures of several millimeters cubed. These non-destructive techniques give access to time-resolved 3D microstructure characterization.

3. Experimental Methods and Materials

One major aim of this thesis is to integrate experimental and modeling efforts in the characterization of microstructure evolution in strontium titanate. The tomography experiments that yield 3D interface networks of real structures and the vertex dynamics model chosen for the growth simulations will be explained in detail in this chapter.

3.1. X-ray Diffraction Contrast Tomography

X-ray Diffraction Contrast Tomography (DCT) imaging is used to characterize the microstructure evolution in strontium titanate in 3D. In this section, the applied technique is described briefly. Detailed descriptions of the experimental setup and the reconstruction procedure are published in [73] and [74].

3.1.1. Principle

DCT is a synchrotron based non-destructive characterization method for polycrystalline materials allowing to investigate their morphology and crystallography in 3D. Using monochromatic coherent high energy X-rays, the technique is a variant of conventional micro tomography [76] reconstructing individual grain shapes from extinction contrast data collected during tomographic scans. During DCT scans, the extinction contrast data is complemented by the simultaneous collection of diffraction information. This additional data is used for the identification of the crystallographic orientation of every single grain with respect to the global reference system as well as the shape reconstruction.

For a given wavelength λ, diffraction information is directly accessible from the constraint that only crystallites fulfilling Bragg's condition

$$2d_{hkl} = \sin\Theta = n\lambda \tag{3.1}$$

will cause constructive interference of the X-rays and therefore a diffraction spot on the detector. Here, d_{hkl} denotes the lattice plane spacing for the set of hkl lattice planes, n is an integer representing the order of reflection and λ marks the wavelength of the X-rays and Θ denotes the angle between the incident ray and the scattering lattice plane. Knowing several projections allows to calculate the hkl lattice plane type for the diffraction event giving access to the crystallography of the particular grain.

3.1.2. Specimen Preparation

The experimental setup of the DCT technique imposes several restrictions on shape and microstructure of the investigated strontium titanate specimens: Rotating tomography specimens should preferably be of a rotational symmetry, minimizing intensity variations arising from sharp edges. Due to the acquisition geometry (presented in chapter 3.1.3) and the limited penetration depth of the achievable X-ray energy, the diameter of the specimen must be below 350 μm. With an effective detector pixel size of 1.4μm and 0.7μm respectively (depending on the applied X-ray optics), grain diameters of 20-30μm are needed to achieve a reasonable resolution at the grain boundaries.

The strontium titanate specimen presented throughout this thesis is cylindrical with a diameter of 250μm, see figure 3.1. It is prepared from raw material processed as described in section 3.3.2. In order to obtain the desired shape the raw material is cut into disks with a thickness of 1mm and a diameter of 3mm using a diamond wire saw. These disks are glued onto tungsten wires of decreasing diameters from 2mm to 0.5mm and ground down to the diameter of the wire using a turning lathe and abrasive paper. During excavation of the last 200-250μm necessary to reach the desired diameter, the specimens rest on the ⌀ 0.5mm tungsten wire.

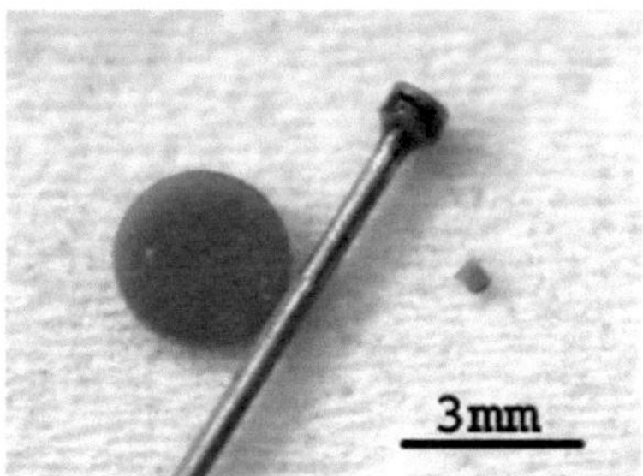

Figure 3.1.: Raw material disk, intermediate shape during abrasion and tomography specimen in its final shape.

3.1.3. DCT Data Acquisition

All presented DCT experiments are performed at the materials research beam line ID11 at the European Synchrotron Radiation Facility (ESRF) in Grenoble, France.

The experimental setup is depicted in figure 3.2: The cylindrical strontium titanate specimen is mounted on a rotating specimen holder and illuminated by high energy X-rays. During 360° scans absorption and diffraction information is collected in 0.05° increments using a FReLon CCD camera with 1024x1024 pixel and 2048x2048 pixel chip, respectively. A total of 7200 images are acquired per scan. The resulting effective pixel sizes are 1.4 μm and 0.7μm. The beam energy for the experiments presented in this thesis is 36keV and the sample detector distance 3.3 mm.

For some samples, separate phase contrast tomography (PCT) datasets are acquired using the same set-up, but a larger specimen detector distance. The resulting edge enhancement due to free space propagation (Fresnel diffraction) increases the visibility of small pores.

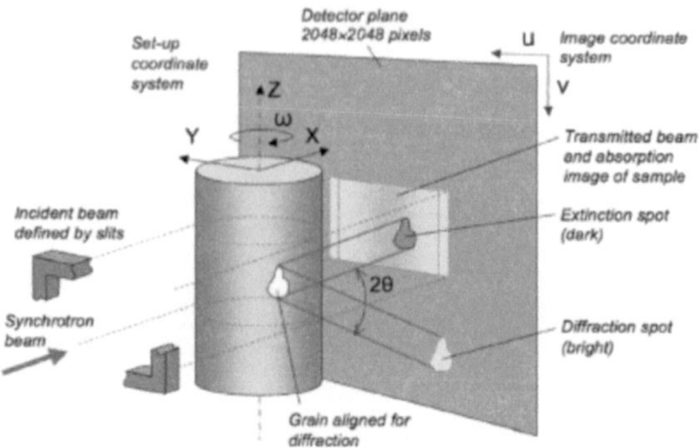

Figure 3.2.: Experimental setup during acquisition of diffraction and absorption information by means of DCT [74].

3.1.4. Microstructure Reconstruction

The applied reconstruction of position and grain shape is primarily based on the exploitation of the diffraction spots gathered during the scans. For the investigated bulk strontium titanate, primarily diffraction spots originating from {110}, {111}, {200} and {220} lattice planes were considered for segmentation. A schematic overview of the reconstruction procedure is shown in figure 3.3. Apart from the algebraic reconstruction, which is carried out using an algebraic tomographic reconstruction algorithm [77], all data and image processing operations were performed using MATLAB 2012a (The MathWorks Inc., Natick, MA, USA).

In a first step, diffraction spots are segmented from the raw images. Due to contrast differences of several orders of magnitude, a double threshold segmentation method is applied: A first threshold is used to identify bright regions in the raw image as possible diffraction spot. The brightest pixel (seed) of each of these regions is stored in a database. A second threshold value is calculated depending on the seed intensity. The region around the seed is then segmented using this value. Different thresholds are thereby used for different diffraction spots depending on the brightness of the individual spot.

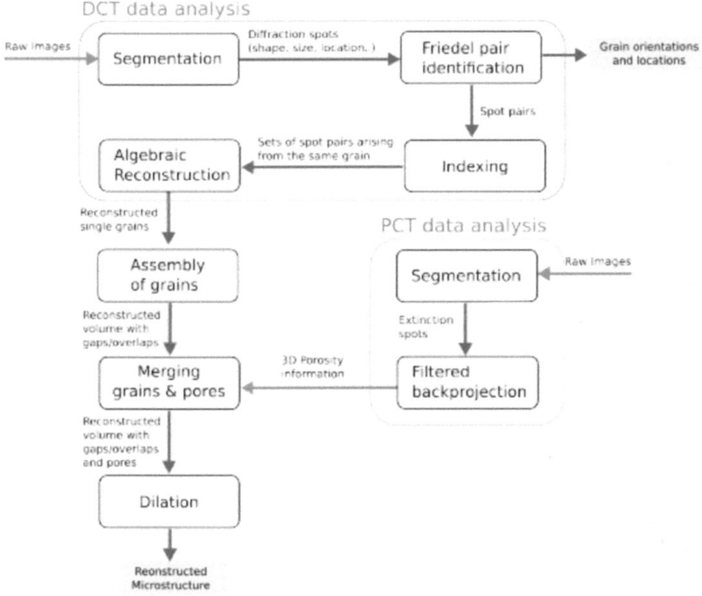

Figure 3.3.: Flow chart of the applied reconstruction procedure.

Information on size, location, shape and intensity of each diffraction spot is stored in a database. Diffraction spot pairs separated by 180° specimen rotation (so called Friedel pairs) are identified. Based on the set-up geometry and the crystallography of the material, the centroid of the diffraction spots and their intensity can be used to calculate the location of the matching spot. If several candidate spots are found, size and shape of the spot are used to identify the correct Friedel pair.

Diffraction angles and the diffraction path through the specimen are calculated for each pair, giving direct access to the crystallographic orientation of the grain which the diffraction spots arise from. Friedel pairs originating from the same grain are collected in an iterative indexing step using spatial and crystallographic criteria [74]. Once at least two Friedel pairs are identified to belong to the same grain, its position inside the specimen is evaluated as intersection point of connecting lines between the diffraction spots forming the pairs.

A standard algebraic reconstruction procedure [78] is applied to these sets of projections, yielding a 3D reconstruction of the particular grain. These reconstructed individual grains are then placed at their correct positions in the specimen resulting in a first reconstructed voxelated grain map that still contains regions of grain overlap and/or unassigned voxels. At this stage, the collective pore ensemble, reconstructed by means of filtered back projection from the PCT data set of the specimen, is subtracted. During post processing ambiguously assigned voxels are zeroed and a uniform dilation procedure that deals with the unassigned voxels is applied. In order to ensure a space filling microstructure reconstruction, grains are not allowed to grow inside other grains or pores nor beyond the specimens volume as obtained from absorption tomography.

3.2. Electron Backscatter Diffraction

2D Electron Backscatter Diffraction (EBSD) measurements on cross sections of the specimen provide a suitable validation of 3D microstructure reconstructions obtained by means of DCT. A quantitative comparison of both techniques requires EBSD measurements on one or more cross

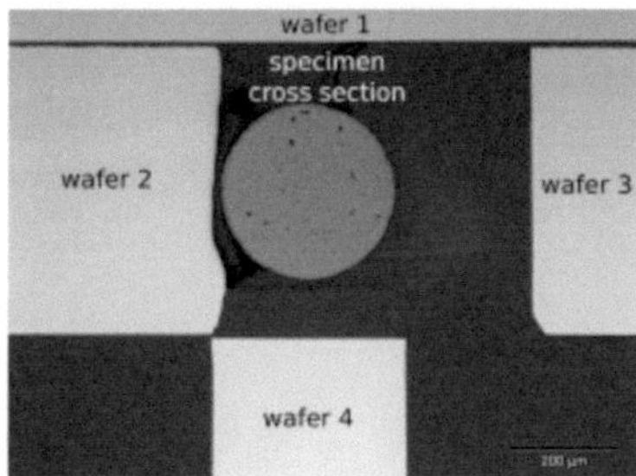

Figure 3.4.: Optical micrograph of specimen and silicon wafers embedded in epoxy resin.

sections of the specimen as well as the identification of corresponding cutting planes in the reconstructed DCT microstructure.

3.2.1. EBSD Measurements

For the EBSD measurements, the specimen is prepared by mechanical grinding and polishing. The surface of the specimen is polished by low angle Argon ion beam sputtering and covered with a thin carbon film to improve electrical conductivity. Subsequently, the specimen is embedded in epoxy resin and surrounded by silicon wafers to improve the electrical conductivity and the mechanical stability, see figure 3.4. One of these silicon wafers is mounted edge on and used for depth measurement and alignment of the specimen. The measurements are performed using a scanning electron microscope (Zeiss; Supra 55 VP) at 15kV equipped with an EBSD system (EDAX TSL). The specimen is mounted with 70° tilt to the EBSD detector in a working distance of 11mm. Data acquisition is performed on the entire cross section in a hexagonal grid with 1 µm step size.

The lateral resolution of the EBSD technique is higher than the chosen step size. It depends on the energy of the primary electrons, the pattern quality [79], the material and the orientation of the specimen [80, 81]. Diffraction patterns are indexed and groups of three or more pixels with

Figure 3.5.: 3D reconstruction and cutting plane that is identified as best match for the EBSD scan. Grains are colored randomly.

a misorientation of less than $3°$ are identified as grains. Grain boundary networks are generated using the *OIM* software package (EDAX Inc., Mahwah, NJ, USA). Post-processing involves scaling of the total image and unidirectional stretching to correct for tilt.

3.2.2. Identification of Corresponding Cross Sections

Due to the manual sectioning process that is performed during the EBSD measurements, the slices through the specimen might be slightly tilted with respect to the cylinder axis. Moreover, the prepared sections might be uneven, so that the resulting images originate from at first unknown cutting planes. Therefore, cross sections through the reconstructed DCT structure that match the EBSD sections need to be identified.

The identification is done using a plane fitting approach based on the spatial distribution of the porosities. Centers of mass of spherical intragranular pores which extend over a few voxels in the DCT reconstruction are especially suited for this technique. Their coordinates are extracted from the EBSD cross sections and manually identified in the reconstructed structure using distinct patterns of neighboring pores. Then, a plane is fitted using the least squares approach. The obtained plane parameters are used to visualize the final slices. Figure 3.5 shows the 3D microstructure reconstruction and one of the identified cutting planes.

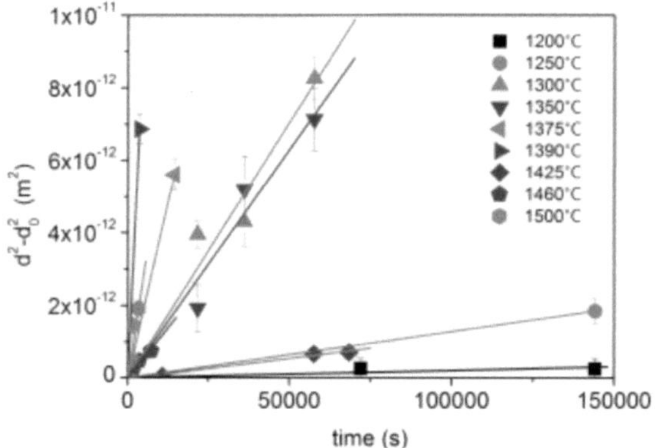

Figure 3.6.: Change in mean grain size of strontium titanate during annealing at various temperatures.

3.3. Strontium Titanate

3.3.1. Properties

Strontium titanate ($SrTiO_3$), an oxide of strontium and titanium, is chosen as the model system for perovskite ceramics because it is stable in the cubic crystal system above $-168°C$. Previous investigations of the material reveal several properties that motivate a further investigation of the link between morphology and grain boundary properties: The material exhibits a strong anisotropy in interface energy [82] and mobility [4] as well as in elastic and plastic behavior. The anisotropy ratio for the elastic constants is 3.5 [83]. Annealing experiments with varied heating conditions [84] reveal a non-Arrhenius temperature dependency of the effective grain boundary mobility, see figures 3.6 and 3.7. For compositions with Sr/Ti ratios of 0.996 and 1.02 and sintering temperature of 1450°C, bimodal grain size distributions are obtained revealing abnormal grain growth [85]. Figure 3.8 shows a SEM image of an abnormally growing grain in $SrTiO_3$. This phenomenon is linked to faceting on {100} planes [86,87] in strontium titanate.

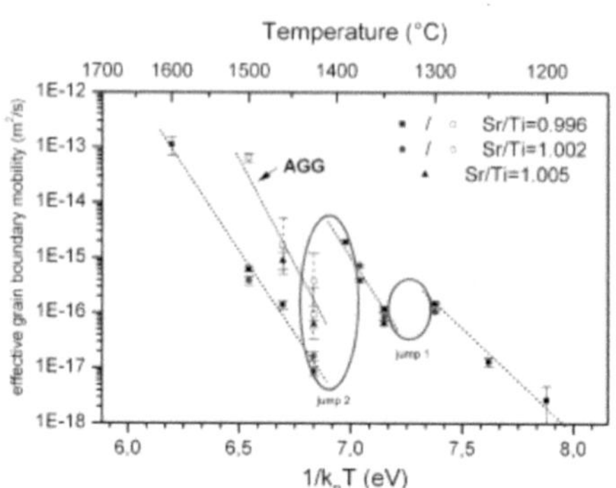

Figure 3.7.: Effective grain boundary mobility as a function of inverse temperature for material with various Sr/Ti ratios [84]. Drops in effective mobility occurring at distinct temperatures are highlighted in red.

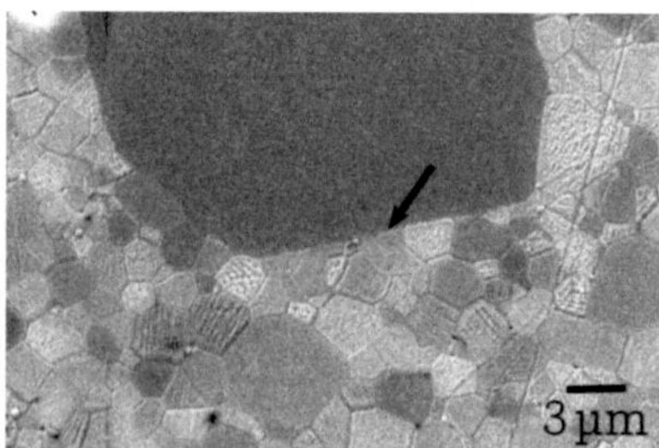

Figure 3.8.: SEM micrograph showing an abnormal grain in $SrTiO_3$ [84].

3.3.2. Processing

The material considered in this thesis is prepared by Michael Bäurer and Wolfgang Rheinheimer at the Institute for Applied Materials Ceramics in Mechanical Engineering [84]. $SrTiO_3$ powders are processed via the mixed oxide route from $SrCO_3$ and TiO_2. Both powders are purchased from Sigma Aldrich Chemie, Taufkirchen, Germany and are 99.9% chemically pure. After undergoing attritor milling for 4h at 1000min^{-1}, the powders are dried and calcined for 6h at $975°C$ undergoing the solid state reaction

$$SrCO_3 + TiO_2 \rightarrow SrTiO_3 + CO_2. \tag{3.2}$$

The powder is uniaxially compacted in a 15mm shaping die at a pressure of 8MPa. The resulting green bodies are then subjected to cold isostatic pressing at 400MPa to reach the final green density. The specimens are sintered and annealed at temperatures in the range of $1300°C$ to $1600°C$ in oxygen atmosphere. To conserve the high temperature status of the microstructure, the specimens are quenched by removing them from the hot zone of the oven after sintering. During this process the specimens remain in oxygen atmosphere.

The results presented in this thesis will focus on the DCT investigation of a high temperature regime specimen that was sintered for 1h at $1600°C$. After the first DCT scan, this specimen is annealed for an additional hour at $1600°C$.

3.3.3. Modeling Parameters

The orientation dependent grain boundary energy for the simulation of normal grain growth in $SrTiO_3$ is assumed to be given by the average of the surface energies of the neighboring grains, see equation 4.11. The surface orientation dependent energy functional for $SrTiO_3$ as obtained by means of capillarity vector reconstruction [82] is employed. Figure 3.9(a) shows the experimentally obtained surface energy values as well as the continuous 2D Fourier series fit employed by Sano *et al.* [82] around the perimeter of the standard stereographic triangle, 3.9(b) shows a stereographic projection of the fitted energy potential.

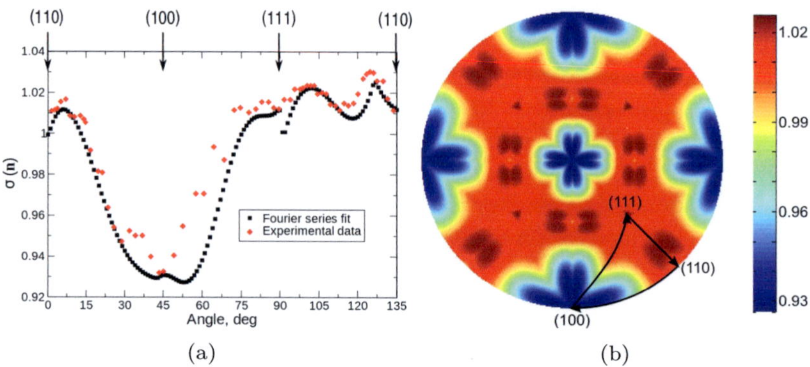

(a)   (b)

Figure 3.9.: Normal dependent surface energies in SrTiO$_3$ [82]: (a) experimentally obtained and Fourier fitted relative surface energies around the perimeter of the standard stereographic triangle; (b) stereographic projection.

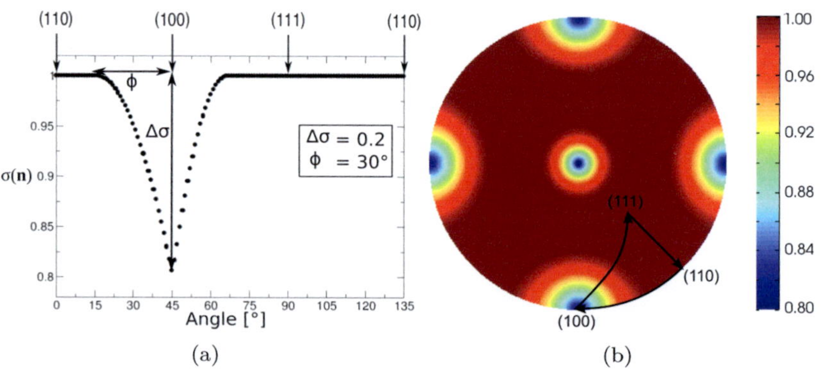

(a)   (b)

Figure 3.10.: As figure 3.9 for smooth one cusp surface energy potential with opening angle $\phi = 30°$ and cusp depth $\Delta\sigma = 0.2$.

Since the evaluation of the grain boundary energy and its derivative is realized in an angular stepping in the vertex dynamics model, the complex shape of the cusps of the energy functional shown in figure 3.9(b) and the resulting abrupt changes in energy may lead to numerical problems. Hence the simplified and probably more realistic potential displayed in figure 3.10 is developed. This potential contains only one smooth cusp located at the minimum energy orientation of the original surface energy potential. The elimination of local minima and discontinuities leads to smoother energy transitions. Showing a sharper $<100>$ cusp than the Fourier series fit given by Sano *et al.*, the smoothed functional is also closer to the experimentally obtained energy values for the lowest energy orientation.

In the context of the present work, variations of this simplified surface energy functional are employed. All applied energy functionals can be characterized by number and shape of the low energy cusps, affecting the overall attraction zone and the energy gradient for orientation variation. Here, the term attraction zone refers to the area of the stereographic projection of the surface energy functional with $\sigma(\mathbf{n}) < 1$. Cusp shape variations are performed by altering the maximum depth $\Delta\sigma$ and the opening angle ϕ of the cusp. Cusp depth $\Delta\sigma$ and opening angle ϕ are marked in the energy potential displayed in figure 3.10(a). The (100) cusp in this energy potential is build according to

$$\sigma(\mathbf{n}) = 1 - \Delta\sigma \left(\frac{\theta}{\phi}\right)^2 \tag{3.3}$$

with $\theta = acos(\mathbf{n}_3)$ and cusp opening angle $\phi = 30°$ and a cusp depth $\Delta\sigma = 0.2$.

Reliable data on orientation dependent relative grain boundary mobility in $SrTiO_3$ is lacking in literature. Experiments investigating a single crystalline front growing into a polycrystalline matrix are reported in [2]. These experiments yield inclination dependent and temperature dependent relative interface mobilities. However, these results are restricted to a handful of distinct orientations and distinct annealing temperatures. Therefore, all simulations including orientation dependent grain boundary mobility anisotropy are performed using artificially designed mobility functionals.

4. Simulation Method

This chapter presents detailed information on the 3D Vertex Dynamics Model used to study the growth kinetics of polycrystalline materials. A brief introduction into the underlying physical principles is followed by a detailed description of the implementation with special emphasis on the model topology, discretization and timestepping. Finally, the extension of the model towards the simulation of systems with orientation dependent grain boundary properties is addressed. Detailed information on rediscretization and topological events during the time evolution of the modeled microstructure can be found in the appendix.

4.1. Vertex Dynamics Model

4.1.1. Physical Principles

The Vertex Dynamics Model applied in this work represents the grain boundaries in a polycrystal as a nodal network. It is based on the assumption, that grain boundary motion can be derived from a minimization of the system's total grain boundary energy

$$V\left(\{\mathbf{r}\}\right) = \int_{GB} \gamma(\mathbf{n}(a))da, \qquad (4.1)$$

where $\gamma(\mathbf{n}(a))$ denotes the grain boundary energy and a is the parametrized grain boundary coordinate.

The energy dissipation arising from grain boundary motion can be derived as [53]

$$R\left(\{\mathbf{r}\},\{\mathbf{v}\}\right) = \frac{1}{2}\int_{GB} \frac{v\left(a\right)^2}{M(a)}da, \qquad (4.2)$$

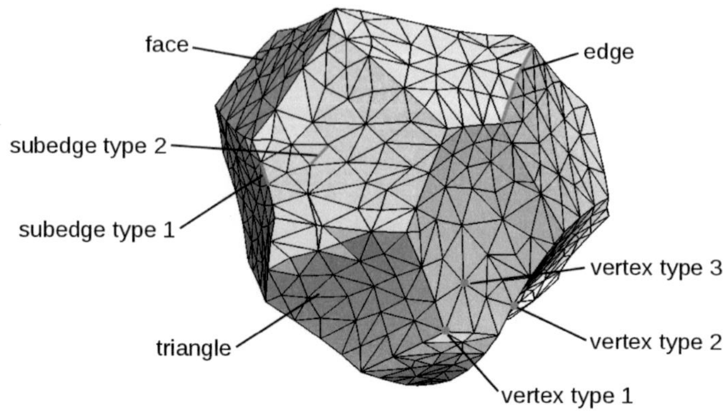

Figure 4.1.: The six elementary geometrical objects described in table 4.1 illustrated on a sample grain.

where $\mathbf{r}$ and $\mathbf{v}$ are the positions and velocities of the grain boundary, $v(a)$ is the velocity component normal to the boundary plane and $M(a)$ is the mobility of the grain boundary at position a.

Then, the equations of motion for the nodes can be derived from the Lagrange equation

$$\frac{\partial R}{\partial \mathbf{v}} = -\frac{\partial V}{\partial \mathbf{r}}. \tag{4.3}$$

4.1.2. Implementation

The implementation of the model is based on the 3D Vertex Dynamics Model published by Weygand [24] with the addition of [54]. The topology of the grain boundary ensemble is composed of six elementary geometrical objects: vertices, subedges, triangles, edges, faces and grains, see table 4.1 and figure 4.1. Vertices and subedges can be classified according to their role in the model that can either be physical (type 1) or discretizational (type 2 and 3, respectively). Discretizational vertices can be divided into vertices discretizing a triple line (type 2) and vertices on a face (type 3).

Vertex	Point in space with distinct position and velocity. Type 1: Physical object. Endpoint of triple lines (quadruple point). Type 2: Discretizational object located on triple line. Type 3: Discretizational object located on face.
Subedge	Straight line connecting two vertices. Type 1: Subedge located along a triple line. Connecting vertices of type 1 or 2. Type 2: Subedge located within a face. At least one of its vertices is of type 3.
Triangle	Basic discretization unit for grain boundaries. Triangles are formed by three subedges.
Edge	Line that is shared by three grains (triple line). Discretized by subedges.
Face	Plane representing a grain boundary. Bounded by edges and discretized by triangles.
Grain	Polyhedron containing at least two faces.

Table 4.1.: Six elementary geometrical objects describing the grain boundary network.

Structure generation and Discretization

Starting structures for the simulations are generated by a randomly seeded Voronoi tesselation of a cube with side length L and periodic boundary conditions in all three spatial directions. These initial grain boundary networks consist of quadruple points and triple lines only and need further discretization in order to allow for smoothly curved grain boundaries. For this reason, additional vertices are introduced both in plane and along the triple lines. This initial discretization is done using a 2D Delaunay triangulation, producing a fairly regular initial mesh.

During the simulation, the triangulation is controlled adaptively based on a critical reference length

$$l_{crit} = \frac{\bar{R}}{n_{virt} + 1},\qquad (4.4)$$

that triggers topological transformations and rediscretization events as a function of the number of discretizational vertices n_{virt} along an edge of the length of an average grain radius $\bar{R}$.

Prior to use in further grain growth simulations, these initial structures are grown assuming isotropic grain boundary energy until they show the properties of a self-similarly growing grain ensemble.

Equations of Motion

A coupled system of equations of motion is derived by expressing the integrals over the grain boundary area in equations 4.1 and 4.2 as sums over the contribution of all discretizational triangles, so that the energy and dissipation potentials are functions of the nodal position $\{\mathbf{r}^A\}$ and velocity $\{\mathbf{v}^A\}$ only.

Using the Lagrange formalism given in equation 4.3 a coupled system of equations of motion for the velocities $\{\mathbf{v}^A\}$ of the N vertices in the system is obtained:

$$\mathbf{D}^A\mathbf{v}^A + \frac{1}{2}\sum_{B,C}^{(A)}\mathbf{D}^{ABC}\left(\mathbf{v}^B + \mathbf{v}^C\right) = \mathbf{f}^A \qquad A,B,C = 1...N \qquad (4.5)$$

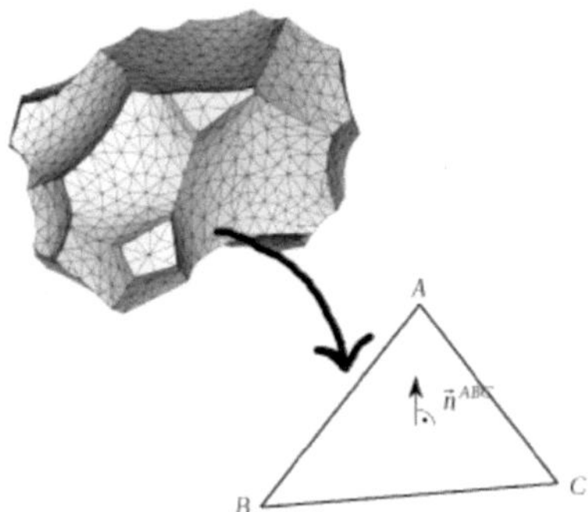

Figure 4.2.: Conventions for the notation of the vertex positions and normal orientation on a basic discretizational unit (triangle).

Here, A, B and C denote the vertices with the positions $\mathbf{r}^A, \mathbf{r}^B$ and $\mathbf{r}^C$ forming a triangle (ABC), see figure 4.2. $\mathbf{D}^A$ and $\mathbf{D}^{ABC}$ are coupling matrices given by

$$\mathbf{D}^A = \sum_{B,C}^{(A)} \mathbf{D}^{ABC}, \tag{4.6}$$

where the sum $\sum_{B,C}^{(A)}$ is defined to go over the vertices B, C of each triangle that is connected to vertex A. The coupling matrix $\mathbf{D}^{ABC}$ is defined as

$$\mathbf{D}^{ABC} = \frac{1}{6M} a^{ABC} \mathbf{n}^{ABC} \otimes \mathbf{n}^{ABC}. \tag{4.7}$$

Here, a^{ABC} is the surface area of the triangle (ABC) and $\mathbf{n}^{ABC}$ denotes its normal vector.

The components of the force vector $\mathbf{f}^A$ are given by

$$f_j^A = \frac{1}{2} \sum_{B,C}^{(A)} \left[\left[\gamma^{ABC} (\mathbf{n}^{ABC} \times (\mathbf{r}^C - \mathbf{r}^B))_j \right] - f_j^{\mathrm{T}} \right]. \tag{4.8}$$

Corresponding to the Herring relation for local mechanical equilibrium at triple lines [88], the last term f_j^T of equation 4.8 represents the torque contributions resulting from inclination dependent grain boundary energies $\gamma(\mathbf{n}(a))$. f_j^T is calculated as

$$f_j^\mathrm{T} = \frac{\partial \gamma^{ABC}}{\partial n_l^{ABC}} R_{li} \frac{\partial n_i^{ABC}}{\partial r_j^A} \tag{4.9}$$

applying the Einstein summation convention for the indices i and l. The rotation matrix $\mathbf{R}$ transforms the change of the normal orientation with respect to the variation of the position of vertex A,

$$\frac{\partial n_i^{ABC}}{\partial r_j^A}$$

which is calculated in the laboratory coordinate system, to the grains coordinate system.

Solving the equation system 4.5 yields the nodal velocities needed for the time evolution of the system.

Dynamics

The time evolution of the model system is implemented using a nested time stepping scheme. A global time stepping, involving recalculation of the complete structure is complemented by a subtime stepping (step 4 of the global scheme) limited to the recalculation of regions with considerable geometrical change.

The overall scheme is structured as follows:

1. start a global time step i

2. calculate geometrical and physical properties of all triangles (ABC)

    a) compute interface normal $\mathbf{n}^{ABC}$ and misorientation Φ

    b) derive grain boundary energy γ, mobility M and nodal forces $\mathbf{f}^A$

3. solve linear system for nodal velocities $\{\mathbf{v}^A\}$

4. enter subincremental scheme

   a) detect critical regions (defined below): calculate a time step Δt^{AB} per subedge

   b) determine the time step Δt given by $\Delta t = \min_{AB}(\Delta t^{AB})$

   c) apply Euler scheme to all vertices: $\mathbf{r}^A(t + \Delta t) = \mathbf{r}^A(t) + \mathbf{v}^A \Delta t$

   d) apply necessary topological transformations to vertices whose velocity has been (re)calculated in the previous (sub)-step

   e) determine vertices, subedges and triangles which need to be recalculated, i.e. those with a short Δt^{AB} or involved in a topological transformation

   f) recalculate geometrical and physical properties of the objects found in step 4(e)

   g) solve linear system for the velocities of the vertices found in step 4(e)

5. check global topology and perform necessary transformations

6. execute global time step $i + 1$ unless stopping criteria are reached.

In step 3 the nodal velocities $\mathbf{v}^A$ are calculated by a conjugate gradient (CG) solver taking the full coupling between vertices into account. As both, the connectivity and the coupling terms may change between time steps, the necessary matrix vector operations for the iterative solver have to be rebuilt after each time step. For flat boundaries the coupling matrix $\mathbf{D}^A$ becomes singular as the in-plane motion of a virtual vertex is frictionless. Therefore a unity matrix with a small prefactor in the range $\kappa = 0.1 \ldots 1$ is added giving an effective friction to the in-plane motion of virtual vertices on a flat grain boundary.

In substep 4(a) the maximal allowed time step of each subedge is calculated based on two conditions: first, a subedge is not allowed to shrink or expand more than a fraction of its own length, e.g. $\Delta l_{max}/l = 0.2$. Second, a subedge is not allowed to rotate out of its initial orientation by more than $\phi_{max} = 2 \ldots 5$ degrees per time step. The values of $\Delta l_{max}/l$ and ϕ_{max} are chosen depending on the inclination dependent grain boundary energy variation.

Significantly varying vertex velocities frequently lead to a very small time step Δt. During sub-timestepping those geometrical objects with a recalculation interval smaller than $n\Delta t$ with n in the range of 5...10 are identified. The environment of these vertices up to their second order neighbors is then considered as critical region and included in the recalculation scheme during step 4(g).

A detailed description of the topological transformations performed during step 4(d) and step 6 is given in appendix A.

The evolution of the grain structure may generate strongly distorted triangles necessitating a local rediscretization. Triangles with inner angles smaller or greater than a user defined critical angle ($15°$ and $135°$ in the current configuration) and subedges longer than a critical length ($1.2l_{crit}$) will be dealt with in a rediscretization procedure depending on the missed quality criteria as described in detail in appendix B.

Since distinct grain boundary energy and mobility values may be assigned to each basic discretization unit, the model allows the simulation of materials with inclination and/or misorientation dependent grain boundary properties.

These scenarios require reliable input data for the orientation dependent grain boundary properties (energy and mobility). For the simulations presented here, all grain boundary energy potentials are based on the assumption that the grain boundary energy $\gamma(\mathbf{n})$ can be approximated from the normal dependent surface energies $\sigma(\mathbf{n})$ of the adjacent grains:

$$\gamma(\mathbf{n}) = \sigma(\mathbf{n})_l + \sigma(\mathbf{n})_r - \sigma_B. \tag{4.10}$$

Here σ_B refers to the binding energy gained when the two surfaces are brought together and new bonds are formed. This binding energy however is often hard to access experimentally. Investigations on MgO suggest the average of the surface energies to be a good description of the grain boundary energy [15, 16]. Accordingly, the normal dependent grain boundary energy $\gamma(\mathbf{n})$ is approximated as:

$$\gamma(\mathbf{n}) = \frac{\sigma(\mathbf{n})_l + \sigma(\mathbf{n})_r}{2}. \tag{4.11}$$

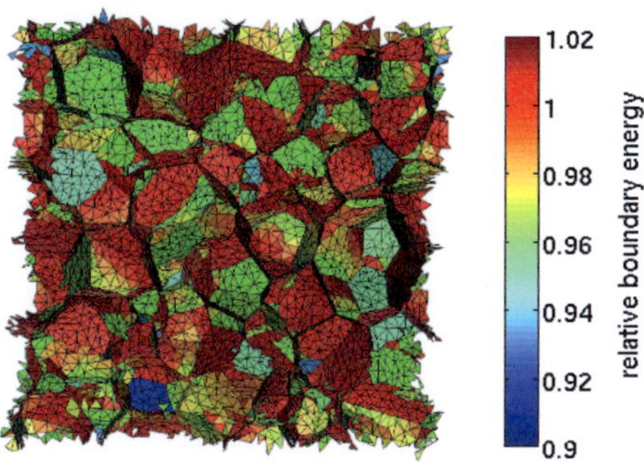

Figure 4.3.: Modeled grain boundary network colored according to boundary normal dependent grain boundary energy.

An example for a triangulated grain boundary ensemble colored according to normal dependent interface energy is given in figure 4.3.

Unfortunately reliable experimental data on relative grain boundary mobility in $SrTiO_3$ is not available. Therefore, orientation dependent grain boundary mobility potentials are built in a similar manner as orientation dependent grain boundary energy potentials, averaging a fictional normal dependent surface mobility of the adjacent grains:

$$M(\mathbf{n}) = \frac{m(\mathbf{n})_l + m(\mathbf{n})_r}{2}. \tag{4.12}$$

5. Results

The investigation of microstructure evolution in the model material is approached by a combination of experiments and simulations. In this chapter results from time-resolved microstructure characterization in 2D and 3D are followed by simulation results on the influence of grain boundary parameter variations on the growth dynamics.

5.1. Experimental Results

5.1.1. DCT Annealing experiments

A cylindrical specimen prepared from the polycrystalline bulk material described in section 3.1.2 is characterized by means of DCT before and after exposing it to 1h ex-situ annealing at 1600°C. Microstructure reconstructions of the specimen before and after annealing are generated. These structures are then aligned and identical subvolumes identified.

General characterization

Three dimensional microstructure reconstructions of the specimen before and after annealing are presented in figure 5.1(a) and (b). The overall shape is identical and surface grains can easily be re-identified by their color which is assigned based on their crystallographic orientation. The second scan is performed on a slightly smaller subvolume of the specimen due to a vertical positioning offset.

The common subvolume contains a total of 849 grains in the initial state. 434 grains are connected with the surface and 415 are bulk grains. During

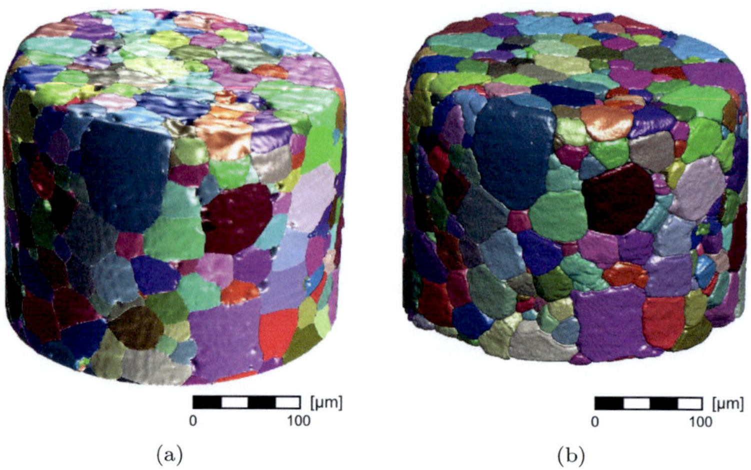

Figure 5.1.: 3D microstructure reconstruction (a) before and (b) after annealing.

annealing, the number of grains shrinks to 797 (398 surface and 399 bulk grains). This corresponds to an average volume growth of 4% per grain. Volumetric grain size investigations reveal that the bulk grains grow from an average grain radius of 14.3±2µm before annealing to 15.0±2µm after annealing. The growth of the surface grains is less pronounced due to surface grooving effects, hindering the free motion of the grain boundaries. Grain size distributions for the bulk grains are given in figure 5.2. Both distributions can be described by a log normal fit

$$f(x) = \frac{1}{x\sigma_S\sqrt{2\pi}} exp\left(-\frac{(lnx - \mu_S)^2}{2\sigma_S^2}\right)$$ (5.1)

with parameters $\sigma_S = 0.42$ and $\mu_S = 0.17$ in the initial state and $\sigma_S = 0.58$ and $\mu_S = 0.20$ in the annealed state. The goodness of fit is tested with a Chi-Square goodness of fit test according to

$$\chi^2 = \sum_{i=1}^{k} \frac{(O_i - E_i)^2}{E_i},$$ (5.2)

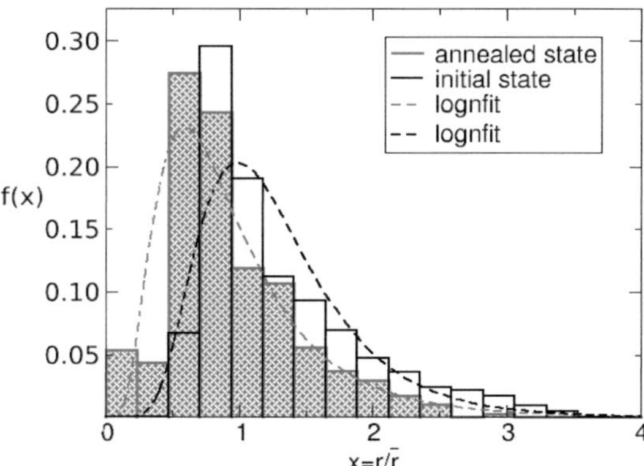

Figure 5.2.: Normalized grain size distributions of bulk grains in initial and annealed state of the tomography specimen. Grain sizes are evaluated from volume data.

where O_i is the observed frequency for bin i and E_i is the expected frequency for bin i. A total of 6771 and 5311 data points are considered. The data is binned with a bin-width of 0.25 resulting in 15 bins. The χ^2 values for the initial and annealed state are $\chi^2=28$ and $\chi^2=22$, respectively.

A comparison of close ups on cross sections taken at the same height of the reconstructed volumes before and after annealing

is shown in figure 5.3. Examples of shrinking and growing grains are encircled with continuous and dashed lines respectively.

Porosity

After sintering, the specimen exhibits a remaining amount of residual porosity which is visible as black areas in figures 5.3(a) and (b). For the initial state, absorption information is used for the reconstruction of pore shapes. The porosity in the annealed state is investigated by means of phase contrast tomography (PCT). Smaller pores are visible in the reconstruction of the annealed state because of the higher sensitivity of the PCT to pores. Accordingly the smallest pores may not have been detected in the initial structure. An example for such a small pore is seen in the intragranular region of the purple colored grain in the upper right region of figure 5.3(b). Figure 5.4 shows an overlay of the porosity before (red) and after (green) annealing. A decrease

Comparison to conventional metallography of bulk material

Figures 5.5(a) and (b) show cross sections through the reconstructed tomography specimen and a scanning electron microscopy (SEM) micrograph taken from the same sintering load of bulk material. Upon visual inspection, similar and realistic grain shapes are identified. A number of intragranular pores are visible in the SEM micrograph, whereas the microstructure reconstruction shows none of these pores. The average grain size of the initial structure is assessed by the equivalent circle diameter (ECD) method: The average grain area is measured and the average grain diameter is calculated from a circle with equivalent area. ECD measurements on cross sections of the reconstructed structure reveal an average grain radius of $13.0 \pm 2\mu$m. The corresponding value obtained by measuring ECD on SEM micrographs of the bulk material shown in figure 5.5(b) is $14.1 \pm 2\mu$m. The average grain radius evaluated from the bulk grains volume approximated as spheres is $14.3 \pm 0.7\mu$m, if surface grains are included one obtains $14.7 \pm 0.7\mu$m.

in pore fraction is visible in the picture. At the same time pore clusters can be re-identified in the annealed state. A calculation of the volume fraction of porosity for both states yields a decrease in volume fraction of porosity from 2.6 ± 0.2 vol-% in the initial state to 1.2 ± 0.1 vol-% in the annealed state. Because of the lower sensitivity of DCT to pores, the initial structure has certainly had a larger volume fraction of pores. Independent measurement of porosity from the sample material in the initial state by the buoyancy method reveals a value of 3.0 ± 0.4 vol-% [2].

A comparison of grain size distributions evaluated by means of ECD on DCT and SEM data is given in figure 5.6(a). The grain size distribution obtained from the volumetric voxel data is given in figure 5.6(b). Although all three distributions can be fitted as log normal distributions, this fit holds best for the distribution obtained from 3D DCT data. The goodness

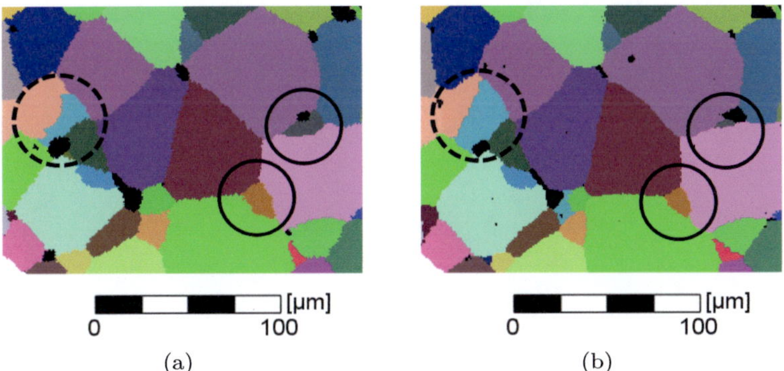

Figure 5.3.: Cross section taken at corresponding height in the reconstructed volume (a) before and (b) after annealing.Examples of shrinking and growing grains are encircled with continuous and dashed lines respectively.

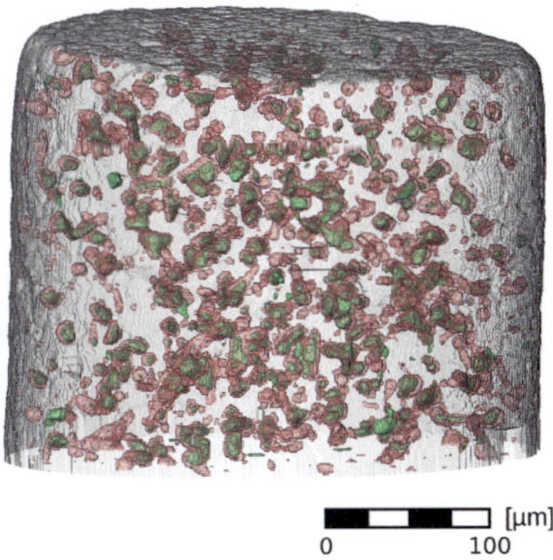

Figure 5.4.: Collective pore ensembles in the microstructure reconstruction before (red) and after (green) annealing.

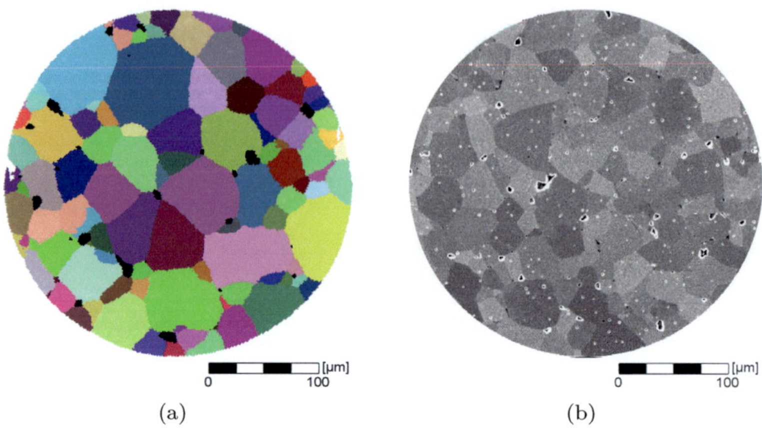

(a)   (b)

Figure 5.5.: (a) Cross section of the reconstructed structure colored according to crystallographic orientation, (b) spherical section of an SEM micrograph of the bulk material.

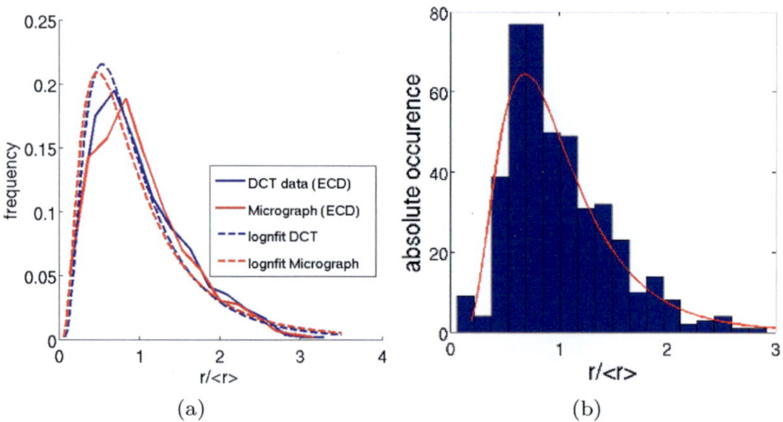

(a)   (b)

Figure 5.6.: Grain size distributions (a) obtained by equivalent circle diameter measurements on bulk DCT data and conventional micrographs, (b) derived from bulk grain volumes in the DCT reconstruction.

of fit, measured as Chi-Square, is 27 for 3D DCT data, 97 for 2D DCT data and 127 for the distribution obtained from micrographs. The fitted log normal distribution for the grain size distribution of the bulk grains evaluated from 3D DCT data has a standard deviation of $\sigma_S = 0.47$.

Interface orientation

Investigations of local grain boundary orientation are performed on individual grains. The grain surfaces are Laplace smoothed and tessellated using the Multiple Material Marching Cubes algorithm described in [89]. Conservation of the physically relevant microstructure elements (triple lines and quadruple points) during smoothing is ensured by restricting the motion of a vertex according to its type: on-face vertices are allowed to move in all directions, on-edge vertices are allowed to move along the edge, while quadruple points are fixed.

Evaluation of the normal orientation of the discretizational elements with respect to the crystallographic orientation of the considered grain for all bulk grains reveals a preference for <100> oriented grain boundaries in the initial state. Figure 5.7 shows a reconstructed bulk grain colored according to the local grain boundary normal orientation. Triangles along triple lines are shown in black. Although this grain consists of 55 faces, its overall shape could almost be described as cube-like with truncated edges showing a clear preference for <100> grain boundary normals.

The preference for <100> oriented grain boundaries observed in the initial state is consolidated post annealing. The frequency of occurrence of local grain boundary orientations is investigated. By discretizing the orientation space in bins of approximately $0.5°$ in both azimuthal and polar angle. Due to the cubic symmetry of the crystal structure, all grain boundary orientations are mapped onto the standard stereographic triangle. The observed distribution is normalized by a random orientation distribution. The corresponding distributions for both annealing states are given in figure 5.8. The distributions show a preference for certain grain boundary orientations reflected in an excess of 15% and 20% with respect to the random distribution.

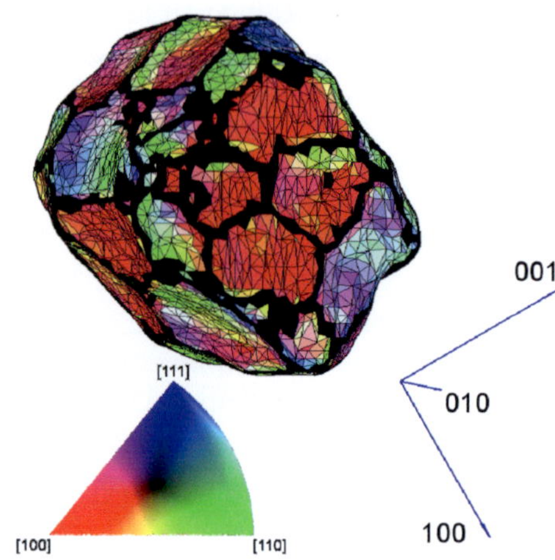

Figure 5.7.: Bulk grain colored according to the local grain boundary normal orientation after surface triangulation. Triangles sharing a triple line are blackened.

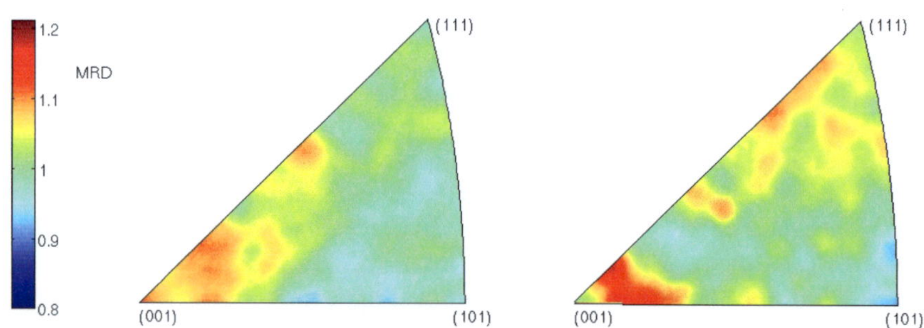

Figure 5.8.: Distribution of local grain boundary orientations for all bulk grains in the initial (left) and annealed (right) state given in multiples of the random distribution (MRD).

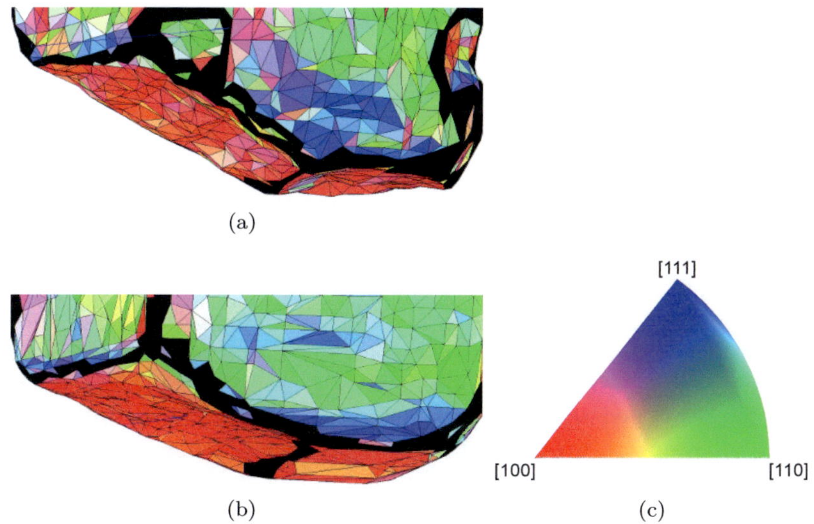

(a)

(b)   (c)

Figure 5.9.: Close up of triangulated grain (representation as in figure 5.7) in the (a) initial and (b) annealed stage. A flattening of faces with normals oriented near [100] is observed.

Upon visual inspection, grain faces, that are mainly oriented near the preferred orientations reveal a tendency to flatten throughout the annealing process. Figure 5.9(a) and (b) show a close-up on two faces that show mainly [100] oriented normals in the initial and annealed state. The big face on the left appears to be more homogeneous in color and thus in grain boundary orientation in the annealed state. The curvature visible in the initial state of the bottom face is diminished after annealing.

Figures 5.10 (a) and (b) show cross sections along the (100)-plane of a large grain and its neighbors before and after annealing. The crystallographic orientations of the neighboring grains are indicated by tripods. The cutting plane is located at the median [100] coordinate. Red and green lines mark the outline of the center grains shape 10% of the grain diameter above and below the cutting plane so as to give an idea about how the cut grain boundary is continued in 3D. From this information it becomes clear, that nearly all cut faces of the pictured grain are oriented parallel to the [100]-axes. Special emphasis shall be drawn to the grain boundaries shared by the central grain and grain #27 and the shared boundary between the central grain and grain #170. These boundaries are also almost parallel to the [010]- and [001]-axes, respectively, indicating a $<100>$ grain boundary normal orientation. Both boundaries appear to be of even more perfect $<100>$ orientation in figure 5.10(b) illustrating the post-annealing state.

Two 3D views of the same grain (#100) are given in figure 5.11, indicating the cutting plane for the cross sections shown in figure 5.10 as black lines. The grain is now colored according to the local migration distance during annealing. The migration distance is determined with reference to a face vertex mesh of the grain in the initial state. A nesting scale of bigger and smaller grains with identical shape is generated in $0.7\mu m$ increments. Every triangle of the post-annealing state mesh is then classified according to which zone of this matryoshka like scale it reached and colored accordingly. From figure 5.11 it is clearly visible, that regions, which are already flat in the initial state exhibit less pronounced changes during annealing as opposed to more corrugated regions, which tend to change a lot.

Misorientation

The grain boundary misorientation angle distribution for the microstructure reconstructions in the initial state is presented in figure 5.12(a). It is found to be very close to the distribution of randomly textured polycrystals [90] which is represented as continuous line.
The misorientation of 3850 grain pairs in the initial and annealed state is compared. Figure 5.12(b) provides the frequency of occurrence of misorientation differences between identical neighbors before and after annealing.

Topological Quantities

The microstructure is further characterized by measuring topological quantities and their evolution during annealing. Surface grains are excluded

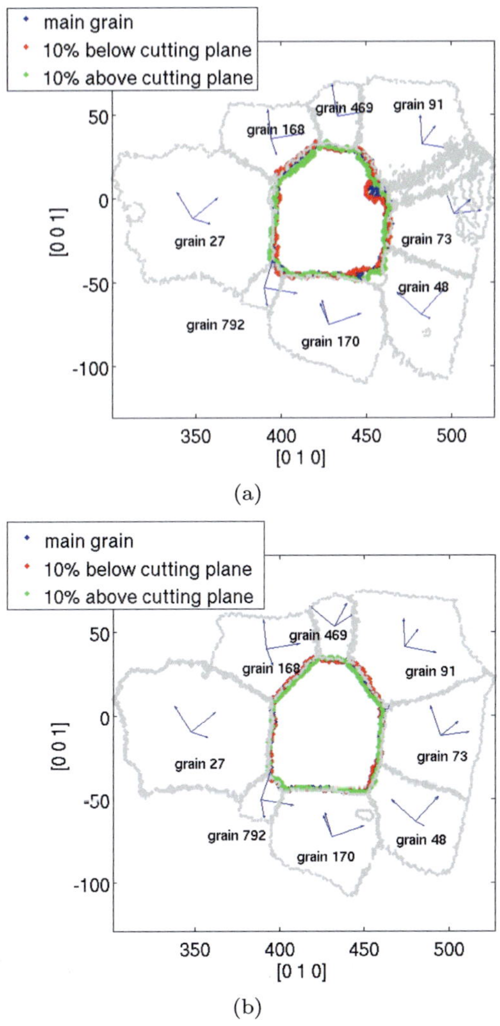

Figure 5.10.: 2D cut along the median (001)-plane of grain #100 and neighboring grains (a) initially and (b) post-annealing. Green and red lines mark the outline of the center grain when cut 10% of the grain diameter above and below the shown cross section.

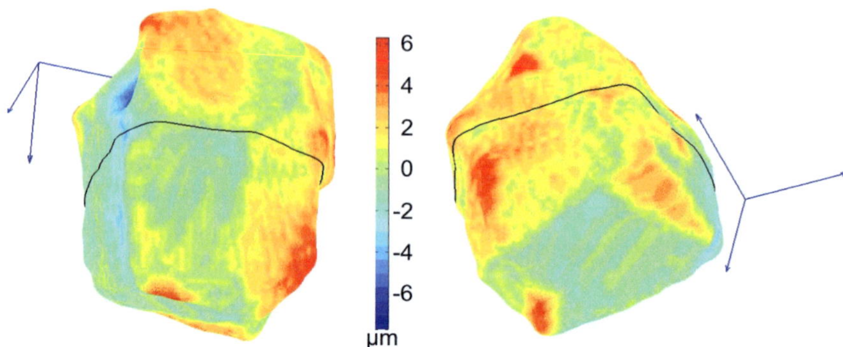

Figure 5.11.: Two 3D views of grain #100 colored according to migration distance during annealing. The black line indicates the cutting plane for the sections shown in figure 5.10.

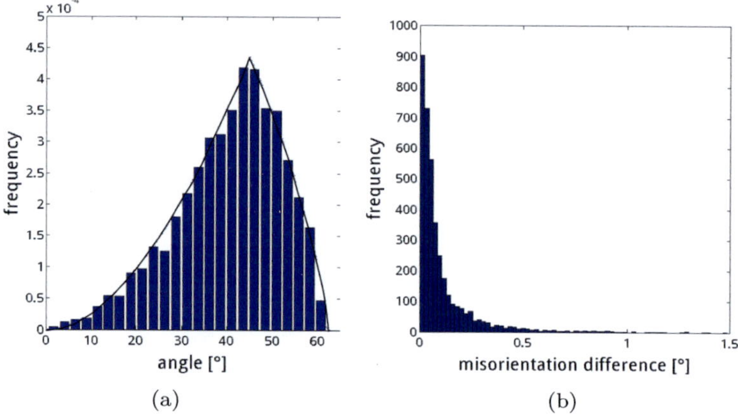

(a)   (b)

Figure 5.12.: Frequency of occurrence for (a) misorientation angles in the initial state and (b) misorientation difference for 3850 identical grain pairs which are present in the initial and annealed state.

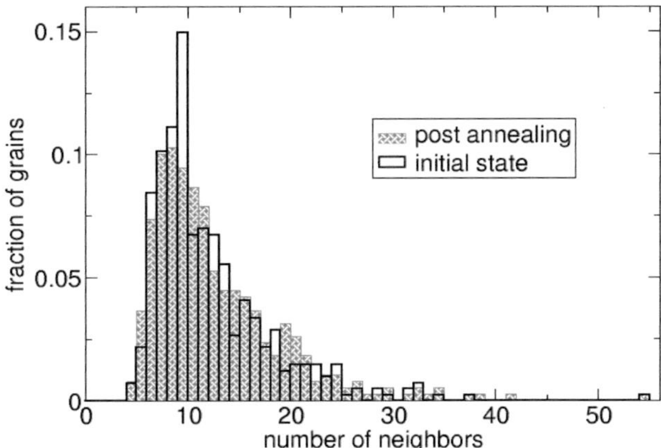

Figure 5.13.: Distributions of number of neighbors per grain for both annealing states.

from this investigation since their shape and thus their topological quantities are altered during specimen fabrication and their growth is hindered by surface grooving effects.

Figure 5.13 shows the distribution of the number of contiguous neighbors (or number of faces per grain) for both annealing states. The distributions look fairly similar, with the mean number of neighbors $\bar{F}_G$ changing from 12.8 in the initial state to 13.3 in the annealed state.

Distributions of the number of edges per grain for both annealing states are given in figure 5.14. Both distributions are best approximated with a lognormal fit with parameters $\sigma_S = 3.25$ and $\mu_S = 0.62$ for the initial and $\sigma_S = 3.29$ and $\mu_S = 0.58$ for the annealed state. The goodness of fit estimated as Chi-Square is $\chi^2 = 49$ for the edge distribution of the initial state and $\chi^2 = 38$ for the annealed state. The distribution does not change significantly during annealing, which is also displayed in the average number $\bar{E}_G$ of 31.1 ± 0.2 edges per grain in the initial state and 32.2 ± 0.2 edges per grain in the annealed state. Both distributions exhibit a long tail caused by the few extremely big grains contained in the structure.

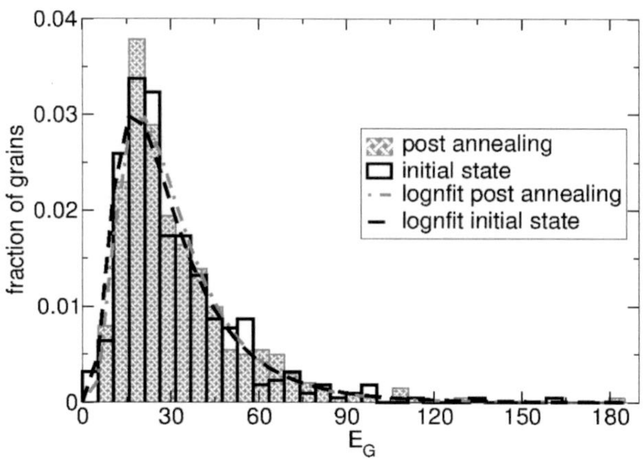

Figure 5.14.: Distributions of number of edges per grain for both annealing states as histogram and log normal fit with parameters $\sigma_S = 3.25, \mu_S = 0.62$ for the initial state and $\sigma_S = 3.29, \mu_S = 0.68$ for the annealed state.

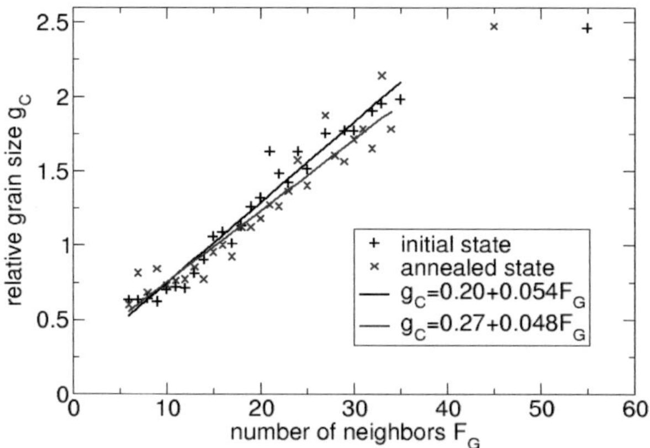

Figure 5.15.: Relative grain size g_C plotted versus number of neighbors F_G for the bulk grains in microstructure reconstructions of the initial and annealed state. Regression lines are given as continuous lines.

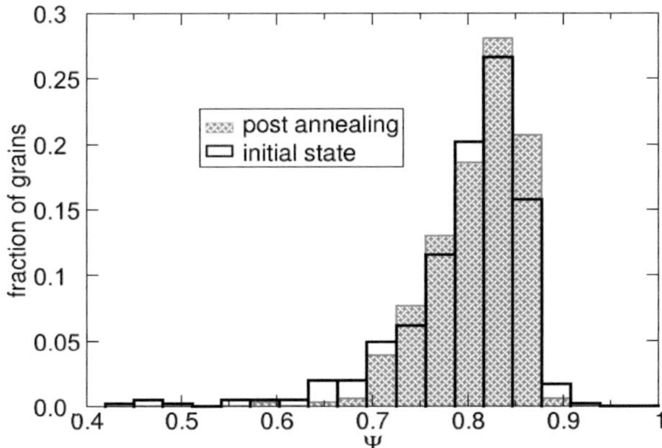

Figure 5.16.: Distributions of sphericity values for both annealing states as histograms.

Figure 5.15 shows the relative average grain size as a function of coordination for bulk grains contained in microstructure reconstructions of the initial as well as the annealed state. The relative grain size is averaged within each class of grains. When the single extraordinary big grain containing 55 grain faces in the initial structure is omitted, a linear correlation is obtained for both annealing states.

Sphericity distributions of both annealing states are compared in figure 5.16. Since calculating sphericity values from voxel data is highly discretization dependent and requires a voxel size that is small compared to the average feature size [91], the sphericity is measured on topology conserving triangulated surface meshes of individual grains according to:

$$\Psi = \frac{\pi^{\frac{1}{3}}(6V_G)^{\frac{2}{3}}}{O_G}.$$

(5.3)

Here, V_G denotes the volume of a grain and O_G the grains surface area. No significant deviation between the sphericity distributions of initial and annealed state is obtained, the average sphericity $\overline{\Psi}$ is 0.80 ± 0.06 before and 0.81 ± 0.15 after annealing. For comparison, the sphericity of a cube is 0.806.

5.1.2. EBSD Validation

A total of eight cross sections are prepared from the tomography specimen presented in section 5.1.1 and characterized by EBSD. All sections are taken in the upper third of the tomography specimen. Best matching cross sections through the DCT reconstruction are identified based on the location of small pores. A detailed description of the applied experimental techniques and the identification process is given in section 3.2.1.

Overlays of the identified DCT

cross sections with the corresponding EBSD networks are presented in figures 5.17(a) and (b). Here, the EBSD data is pictured as grain boundary networks, while the colored grains mark the DCT reconstruction. The spacing between the two EBSD sections along the axis of the cylindrical specimen is approximately 3 µm.

Microstructure Characteristics

The grain boundary networks in figure 5.17(a) contain 107 (DCT) and 108 grains (EBSD) respectively. The average grain radius for these sections is measured by the linear intersect method and is 15.1 ± 0.4 µm (DCT) and 14.9 ± 0.4 µm (EBSD). The grain boundary networks in figure 5.17(b) contain 109 (DCT) and 110 grains (EBSD) respectively. The average grain radius for these sections is 15.8 ± 0.6 µm (DCT) and 14.5 ± 0.5 µm (EBSD). The difference between the average grain sizes as obtained by the different characterization methods is much larger for the second section. Possible explanations for this phenomenon are discussed in chapter 6.

Visual comparison of corresponding slices reveals a good agreement in grain shape and pore size but a higher amount of pores in the EBSD data. The number of pores in the DCT slices reaches approximately 80% to 90% of the pore count in the corresponding EBSD section. The remaining pores, which are missing in the DCT reconstruction are assumed to be very small intragranular pores. Only some of these very small pores (i.e. covering an area smaller than 10 μm^2 in EBSD data) could also be detected in the DCT data set. An example are the small pores in the purple grain in the center image of figure 5.19(f).

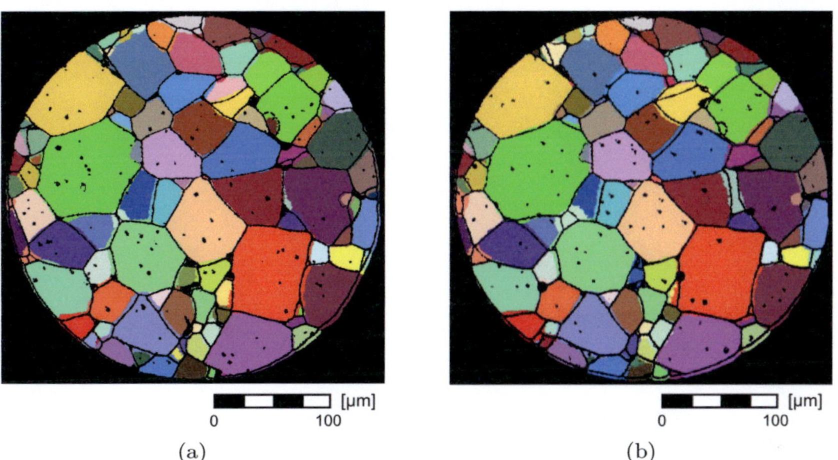

Figure 5.17.: Two different cross sections at height 308 µm and 311 µm in the DCT reconstruction (colored) layered with corresponding EBSD networks.

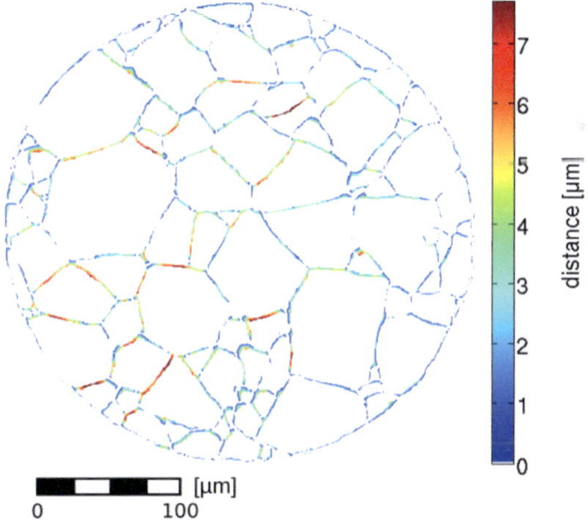

Figure 5.18.: DCT network shown in figure 5.17(a) colored according to euclidean distance to corresponding EBSD network.

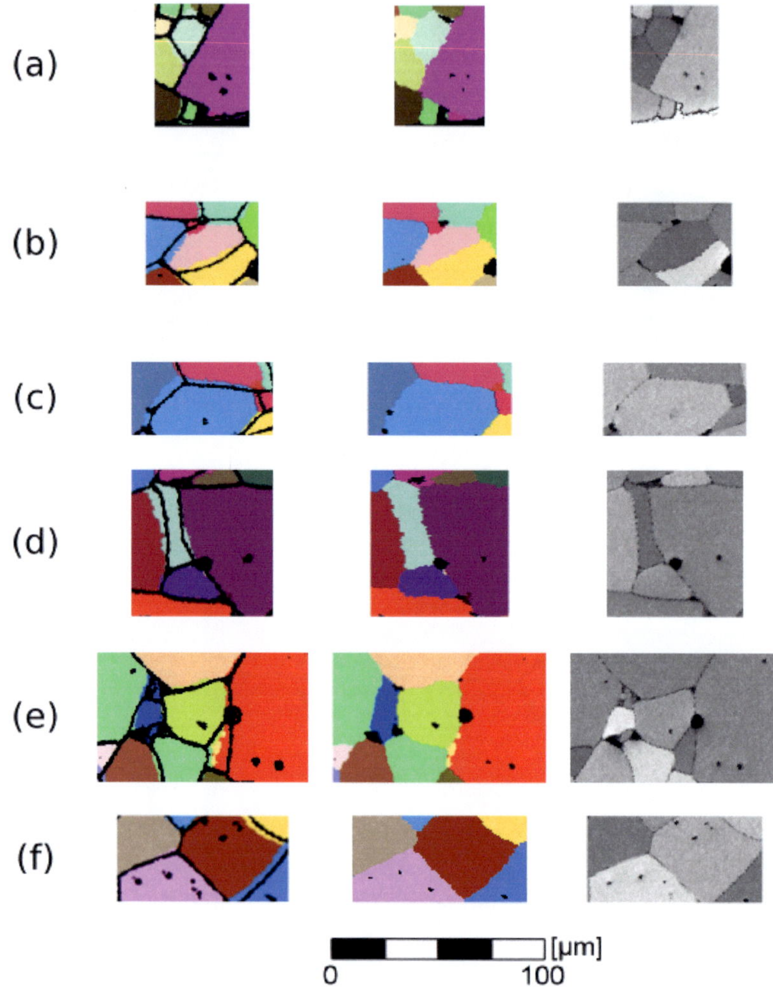

Figure 5.19.: (a)-(f) Close ups of several regions selected from figure 5.17(a) and 5.17(b). DCT with superposed EBSD grain boundary network (left), DCT (center), EBSD greyscale (right)

Euclidean Distance Mapping

The difference between the corresponding DCT and EBSD grain boundary networks is assessed using a 2D Euclidean distance transform [92]. The euclidean distance between each point on the EBSD network and the nearest point on the corresponding DCT network is calculated. The average distance between the corresponding networks is found to be 1.98 µm for the cross section shown in 5.17(a) and 1.95 µm for the cross section shown in 5.17(b). Removing intragranular pores, whose uncertainty is not affecting the accuracy at the grain boundaries, changes the average error values to 1.86 µm and 1.88 µm respectively. The DCT network of the cross section shown in figure 5.17(a) colored according to the euclidean distance to the EBSD network is given in figure 5.18. From this image, it can be seen, that the grain boundaries in the DCT reconstruction appear to be slightly more curved, since the greatest deviation between DCT and EBSD networks is observed on the grain boundaries rather than at the triple points.

Qualitative comparison

Figure 5.19 shows close ups of regions from both inspected cross sections. The superposed representation of DCT grain map and EBSD grain boundary network is complemented by both the DCT (colored) and EBSD (greyscale) grain maps to allow for a more detailed investigation. While corresponding intragranular pores are found in many cases (figure 5.19(a), (d-f)), there are also a few examples for pores revealed in the EBSD analysis, which are not resolved in the DCT reconstruction (e.g. missing pores in the brown grain in figure 5.19(b), the blue grain in 5.19(c), the upper left green grain in 5.19(e) or the brown grain in 5.19(f)). Moreover, the image reveals some pores at doubtable locations in the DCT reconstruction (e.g. the pore inside the pink grain in the upper part of 5.19(d)). Both, the EBSD grain map and the shape of the pore suggest it to be intergranular rather than intragranular.

Figure 5.19(e) reveals a region of very small grains in the EBSD grain map, that is not resolved in DCT (blue region in the left part of the image). Furthermore, a local deviation in curvature is observed. In these cases, the grain boundaries appear to be more straight in the EBSD grain maps than in the DCT maps (e.g. between blue and pink grain in 5.19 (b), in between the blue grains in 5.19(c)) while figure 5.19(a) shows that the surface contour of the two grain maps is nearly identical.

Lastly, a pixel wise smearing presumably resulting from the interpolation during plane fitting is obtained at the grain boundaries in some of the DCT reconstructions. These artifacts occur typically as one voxel wide extrusions reaching two or three voxel lengths into the neighboring grain (figure 5.19(b),(d),(e)).

Orientation Information

In order to identify corresponding grains in both characterizations, the global rotation between the laboratory systems of the different methods had to be identified. Therefore, 24 grains which could be related to each other were determined visually in cross sections obtained by both characterization methods. The identification of the smallest misorientation between the orientation of these 24 crystallites reveals a consistent transformation between EBSD and DCT orientation space for all crystallites. The transformation is a rotation of $20.83 \pm 1.54°$ around the misorientation axis $[0.0461\ 0.9935\ 0.1040] \pm 3°$.

5.2. Simulation Results

Grain growth simulations are a valuable addition to conventional microstructure characterization because they allow to study parameter variations that are not possible experimentally. Three major simulation scenarios are presented throughout this section: two set-ups studying grain growth under isotropic and anisotropic conditions and a set-up investigating the nucleation of abnormal grain growth as a function of relative grain size and relative grain boundary properties.

5.2.1. Grain Growth

Initial configurations for all simulations are grain structures from Voronoi tessellations, that were isotropically coarsened to a regime with self similar grain size distributions, as described in section 4.1.2. The simulation box is periodic in all three dimensions. The grain boundary energy γ and mobility M are treated as dimensionless quantities. If not stated otherwise, the simulations are started from configurations containing 515 grains. The coarsening process is investigated until a lower limit of 100 remaining grains is reached.

For all scenarios, the growth dynamics is assessed as area evolution. Assuming linear growth, the normalized growth rate k is defined as

$$k = \frac{\alpha \gamma M}{A(0)},$$ (5.4)

see equation 2.14. It can be established as the slope of a linear fit to the area evolution obtained from the simulation. For comparison of various grain boundary property anisotropy scenarios a relative growth rate K

$$K = \frac{k}{k_0}$$ (5.5)

is defined, where k_0 is the normalized growth rate of a system with isotropic grain boundary properties.

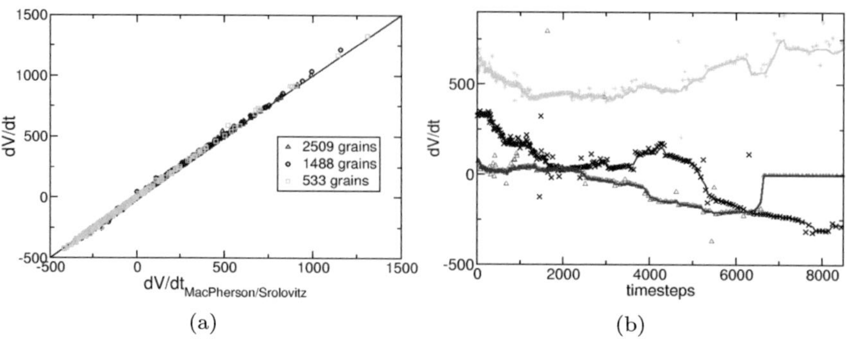

Figure 5.20.: (a) Volume change rates dV/dt of all grains obtained at three stages of a simulation started with 2522 grains plotted against the analytically derived value [33]. (b) Time evolution of the volume change rates obtained from simulation (symbols) and analytical calculation for three randomly chosen grains (continuous line).

Isotropic systems

For the isotropic scenario, the modeling accuracy is tested by comparison with the MacPherson Srolovitz relation 2.19. Initially, the investigated system contains 2522 grains. Grain growth is simulated until the number of remaining grains reaches a lower limit of 500 grains.

Figure 5.20(a) shows the volume change rate of each grain in the system at three steps during the simulation plotted against the analytically derived value according to equation 2.19. Almost all data points lie on a straight line with slope one indicating a very good agreement between analytical and simulation results. Exemplarily, volume change rates of randomly chosen grains showing growing, shrinking and undecided behavior are compared to the analytically derived rates over the course of the simulation in figure 5.20(b). Here, the analytically derived values are plotted as continuous lines, while the symbols mark the volume change rate as obtained from the simulation. The overall agreement is again very good. Occasional differences can be traced to topological changes and rediscretization events in the simulation.

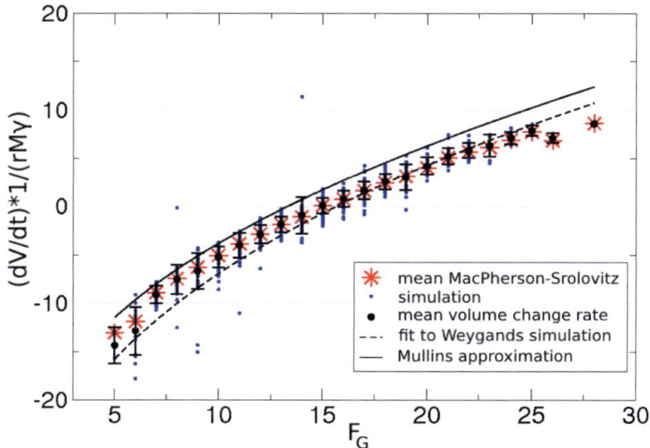

Figure 5.21.: Normalized volume change rate plotted against number of grain faces. Blue dots mark the normalized volume change rate obtained from the simulation, black dots their average for a constant number of grain faces. The mean normalized volume change rate according to MacPherson *et al.* [33] for each class of grains is marked as red stars. Analytical approaches of Mullins [93] and the best fit to Weygands simulation [24] are given as solid and dashed line, respectively.

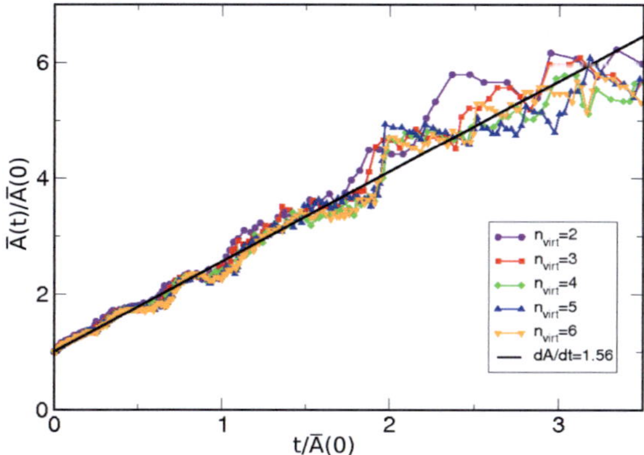

Figure 5.22.: Area evolution as a function of normalized time depending on the discretization.

In figure 5.21 the analytical approximation of Mullins [93] for the normalized volume change rate is plotted against the number of faces F_G together with the normalized volume change rate obtained from the simulation and the best fit to the results presented in [24]. A moderate scatter around the mean values of the simulation is obtained. The volume change rate changes sign for grains with 15 faces, whereas for the basic 3D vertex model [24] a slightly higher value of 15.6 faces is observed. The mean values of both the current vertex model and the analytical expression according to MacPherson-Srolovitz agree, independent of the number of faces as expected from the results shown in figures 5.20(a) and (b) and contrary to the observations reported in [94]. The obtained deviations in predicted volume change rate increase with decreasing coordination number. For very small grains, the results of the new vertex dynamics model are considerably closer to the analytical expression.

A systematic variation of the number of discretizational vertices ($n_{virt} = 2 \ldots 6$) introduced between two physically relevant vertices along a triple line shows that the growth dynamics, expressed as an equivalent area time evolution $A(t)$ is linear in time and independent of the level of discretization, see figure 5.22. Within the range of the varied discretizations the grain size distributions remain unaffected. Therefor, and to allow for anisotropy induced local grain boundary orientation gradients, all subsequent simulations are performed with $n_{virt} = 5$.

Anisotropic systems

The simulations featuring anisotropic grain boundary properties are divided into three major scenarios investigating the influence of (I) grain boundary energy anisotropy, (II) grain boundary mobility anisotropy and (III) combined energy and mobility anisotropy on the growth dynamics.

I. Energy Variation

Simulations studying the influence of grain boundary energy anisotropy on the overall growth dynamics are performed using energy functionals based on the smooth energy functional described in section 3.3.3. The applied

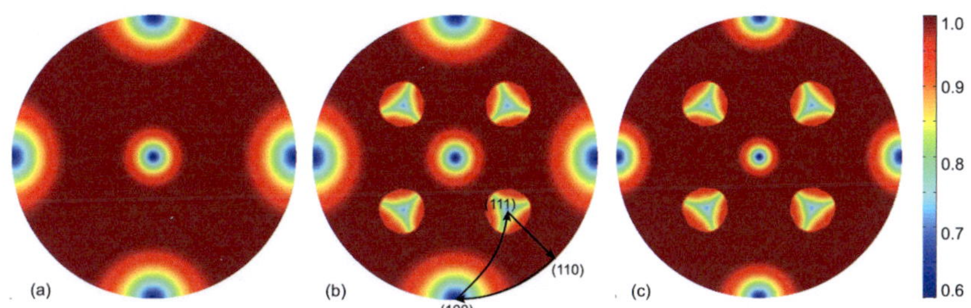

Figure 5.23.: Stereographic projections of the applied energy functionals: (a) simplified one cusp energy functional; (b) as (a) with additional cusp; (c) two-cusp functional with overall attraction zone equal to (a).

functionals differ in number, depth and width of the energy minima. Four major energy variation scenarios are distinguished:

> **(a) effects of torque contribution:** simulations applying the basic normal dependent (one cusp) functional with and without the torque contribution (second term in equation 2.9 and equation 4.8),

> **(b) variation of cusp depth:** simulations applying the energy functional of (a) with torque contribution and varied cusp depth $\Delta\sigma$,

> **(c) variation of cusp width:** simulations applying the energy functional of (a) with torque contribution and varied cusp opening angle ϕ,

> **(d) effects of additional energy minima:** simulations applying energy functionals containing two cusps with torque contribution and varied opening angle of the $<100>$ cusp.

Stereographic projections of the energy functionals are given in figure 5.23. The predominant difference between these functionals is the attraction zone of their energy cusps, where the term attraction zone refers to the area of the stereographic projection with $\sigma < 1$. Accordingly, the half-depth attraction zone is the area of the surface energy functional with $\sigma < (1 - \frac{1}{2}\Delta\sigma)$.

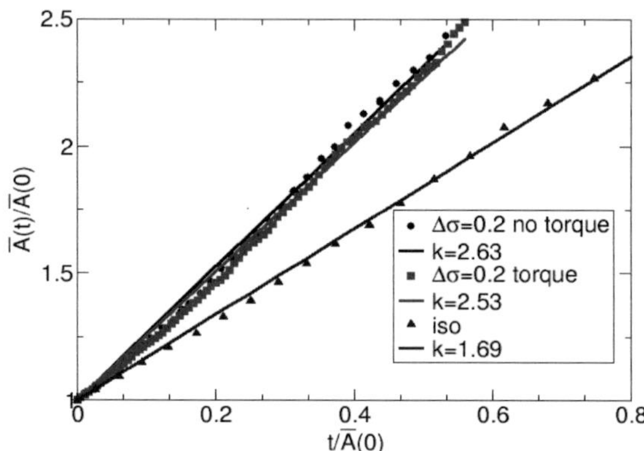

Figure 5.24.: Area evolution as a function of normalized time for simulations with isotropic and anisotropic grain boundary energy with and without torque contribution.

I(a) effects of torque contribution   The growth dynamics for identical starting configurations treated with isotropic and anisotropic grain boundary energy according to the smooth surface energy functional presented in section 3.3.3 are given in figure 5.24. The configuration containing 2500 grains is simulated with a cusp depth $\Delta\sigma = 0.2$ and cusp opening angle $\phi = 30°$ with and without taking into account the torque term in equation 4.8. Both anisotropic simulations exhibit increased growth rates with respect to the isotropic one. The relative growth rate K is 1.50 ± 0.05 for anisotropic systems with and 1.56 ± 0.04 without torque contribution.

Figure 5.25 shows the fraction f_C and $f_{1/2C}$ of triangles whose grain boundary normal with respect to one of the adjacent grains lies within the attraction zone and the half-depth attraction zone of the energy cusp plotted against the number of remaining grains. While f_C and $f_{1/2C}$ remain roughly constant for the simulations neglecting the torque term, a significant increase of in-cusp orientations is observed when considering torque contributions. Accordingly, the average grain boundary energy stays constant for systems not accounting for torque contributions, while a transition to a lower average grain boundary energy is observed when

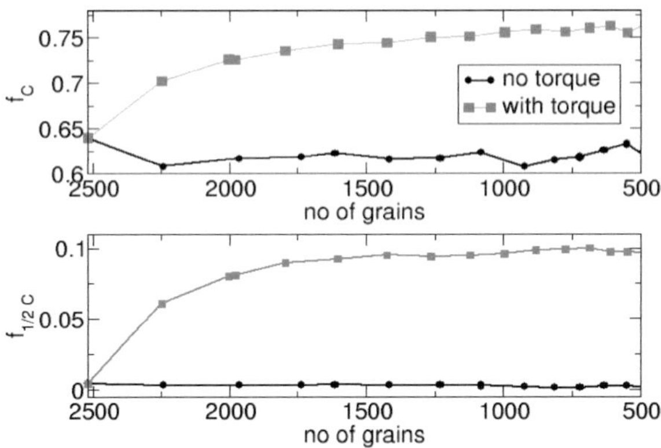

Figure 5.25.: Fractions f_C and $f_{1/2C}$ of discretizational units with grain boundary normal (with respect to one of the two adjacent grains) falling in the attraction zone and the half-depth attraction zone of the energy cusp plotted against number of grains remaining in the system.

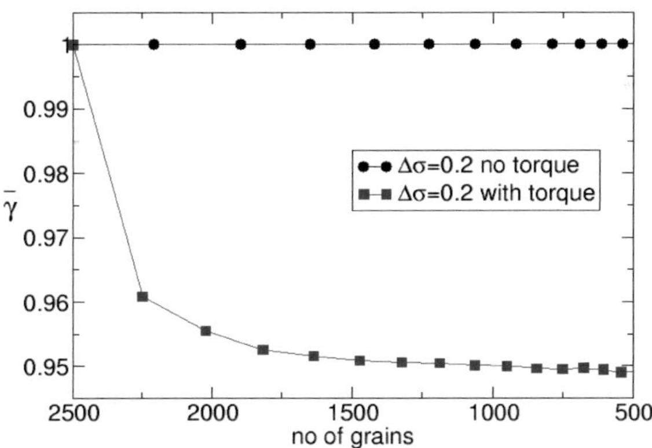

Figure 5.26.: Average grain boundary energy $\bar{\gamma}$ over number of remaining grains for systems with and without torque contribution.

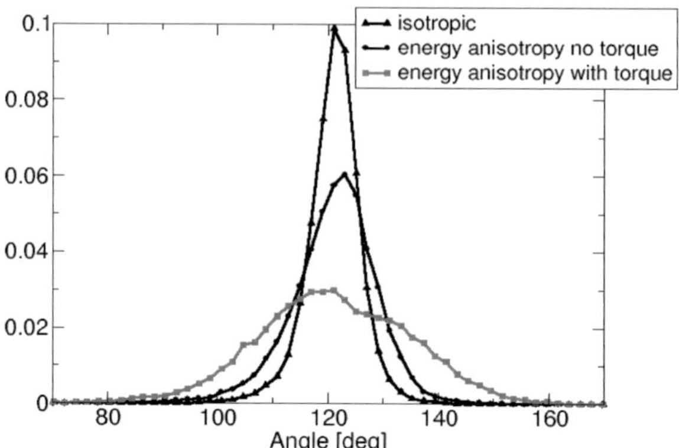

Figure 5.27.: Normalized equilibrium angle distribution for simulations with isotropic and orientation dependent grain boundary energy with and without torque contribution.

accounting for torque contributions, see figure 5.26. The transition to the lower level is completed after about one third of the grains are consumed.

Figure 5.27 shows a comparison of the equilibrium angle distributions for all three configurations, measured after coarsening the structure to half the initial number of grains. A broadening of the distribution is obtained for the anisotropic simulation scenarios. The broadening is more distinct for the configuration accounting for torque contributions.

The evolution away from an isotropic structure dominated by 120° equilibrium angles can also be visualized at the evolving shape of a growing grain: figures 5.28(a) and (b) show a randomly chosen grain in its initial (green surface) and evolved (wireframe) state alongside a tripod indicating its crystallographic orientation. A tilting of certain grain boundaries towards <100> oriented normals (indicated by the angle between the continuous and the dashed line in figures 5.28(a) and (b)) as well as differences in migration distance of various facets are clearly visible.

Normalized grain size distributions of the isotropically and both anisotropically coarsened structures and the corresponding log normal fits are shown in figure 5.29. The grain size r is assessed as radius of a sphere with

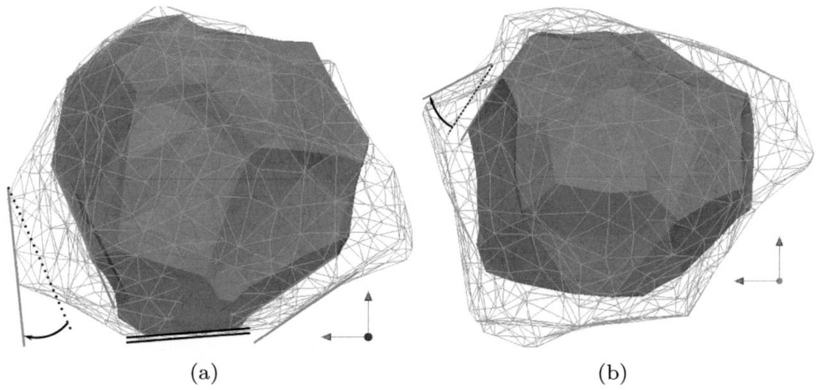

(a)   (b)

Figure 5.28.: Two views (a) and (b) of a growing grain at two stages during anisotropic grain growth simulation with orientation dependent grain boundary energy. The gray surface marks the grain in its initial state, the wireframe shows its evolved state. The grains crystallographic orientation is indicated by a tripod. Exemplarily, two grain faces that tilt throughout the simulation are marked with a dashed line for their initial and a continuous line for their final position.

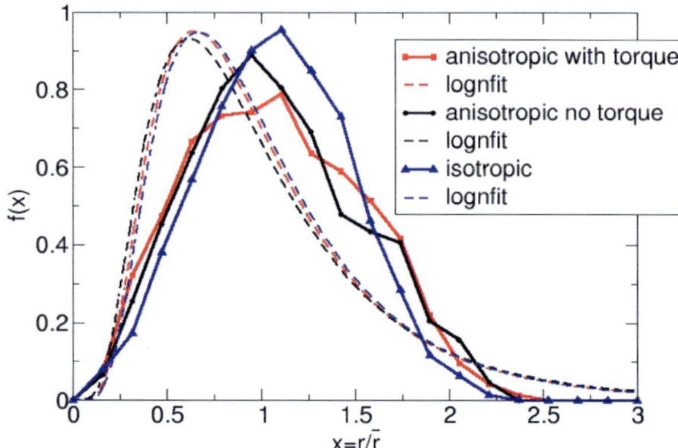

Figure 5.29.: Normalized grain size distributions f(x) as obtained from simulations with isotropic and anisotropic grain boundary energy with and without torque contribution.

equivalent volume. The distributions can not be fitted to a log normal function. For both cases of anisotropic coarsening a broadening of the distribution is obtained. At the same time, the maxima of both anisotropic distributions are shifted towards smaller relative grain sizes.

I(b) variation of cusp depth   The influence of the cusp depth $\Delta\sigma$ on the overall growth dynamics is studied in simulations performed with energy functionals with $\Delta\sigma = 0.05, \dots, 0.2$ and constant opening angle $\phi = 30°$. The resulting relative surface energy around the perimeter of the standard stereographic triangle is given in figure 5.30.

Figure 5.31 shows the fraction f_C and $f_{1/2C}$ of in-cusp orientations over the course of the simulation for the four configurations. The fraction of triangles with low energy orientations increases for all configurations, resulting in a decreasing average grain boundary energy $\overline{\gamma}$, see figure 5.32. Here, the average energy for the four configurations is plotted against the number of remaining grains. After a transition period the average grain boundary energy settles at a new stable value for each configuration. This value is $\overline{\gamma} = 0.995 \pm 0.01$ for the configuration with $\Delta\sigma = 0.05$, $\overline{\gamma} = 0.984 \pm 0.01$ for $\Delta\sigma = 0.1$, $\overline{\gamma} = 0.967 \pm 0.01$ for $\Delta\sigma = 0.15$ and

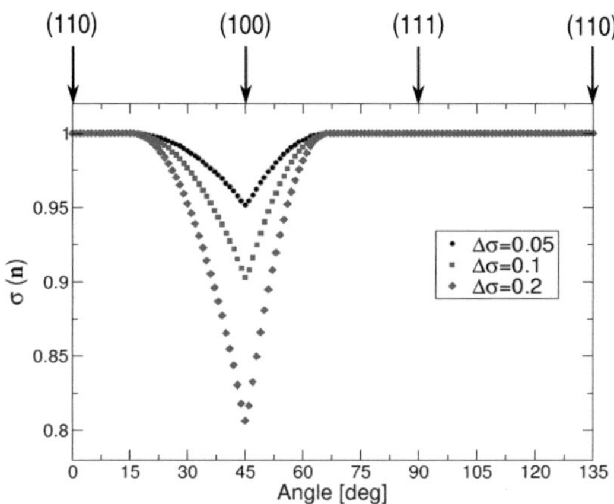

Figure 5.30.: Relative surface energy $\sigma(\mathbf{n})$ around the perimeter of the unit triangle for variation of energy functional cusp depth $\Delta\sigma$.

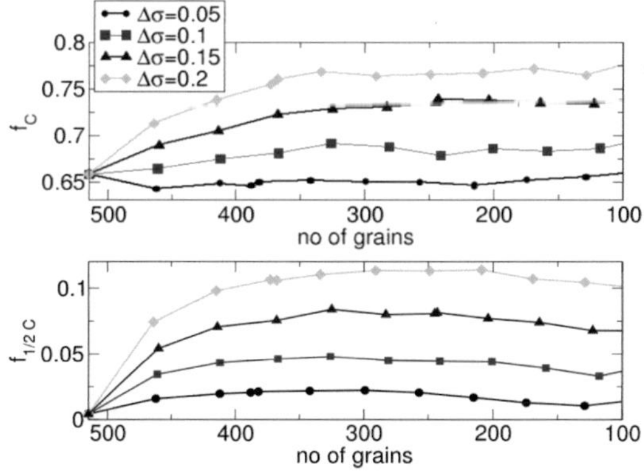

Figure 5.31.: As figure 5.25 for configurations with varied cusp depth $\Delta\sigma$.

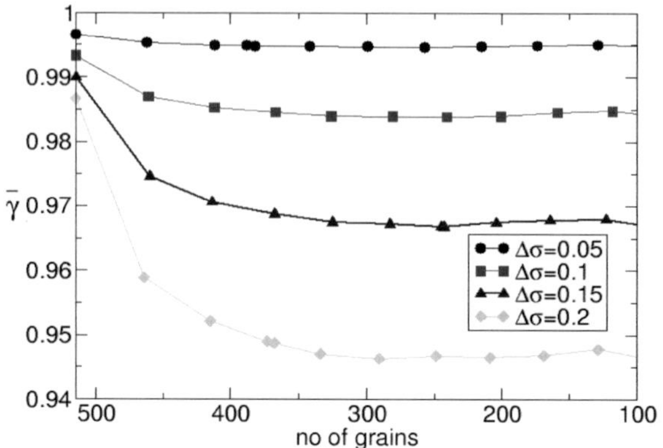

Figure 5.32.: As figure 5.26 for systems with varied cusp depth $\Delta\sigma$.

$\overline{\gamma} = 0.946 \pm 0.01$ for $\Delta\sigma = 0.2$. The relationship between average grain boundary energy $\overline{\gamma}$ and cusp depth $\Delta\sigma$ can be approximated by a linear function

$$\overline{\gamma} = 1 - 0.24\Delta\sigma, \tag{5.6}$$

with a standard deviation of $\sigma_S = 0.002$ for the opening angle $\phi = 30°$.

All anisotropic configurations are found to exhibit accelerated growth rates compared to the isotropic one. Figure 5.33 shows the relative growth rate K as a function of cusp depth $\Delta\sigma$. While a linear increase in relative growth rate with rising cusp depth is observed for the functionals with $\Delta\sigma$=0.05 and $\Delta\sigma$=0.1, the increase in relative growth rate slows down for the deeper cusps.

I(c) variation of cusp width The influence of the cusp width on the overall growth dynamics is studied in simulation scenarios with varied cusp opening angle $\phi = 10°, \ldots, 40°$. For these scenarios, the cusp depth is kept constant at $\Delta\sigma = 0.2$. Figure 5.34 shows the resulting relative grain boundary energy around the perimeter of the standard stereographic triangle.

Figure 5.35 shows the fraction f_C of triangles affected by the attraction zone and the fraction $f_{1/2C}$ affected by the half-depth attraction zone of

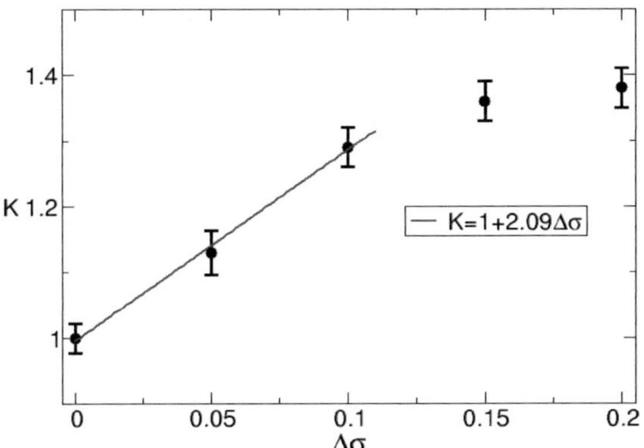

Figure 5.33.: Relative growth rate K plotted against cusp depth $\Delta\sigma = 0, \ldots, 0.2$. Datapoints with error bars and linear regression.

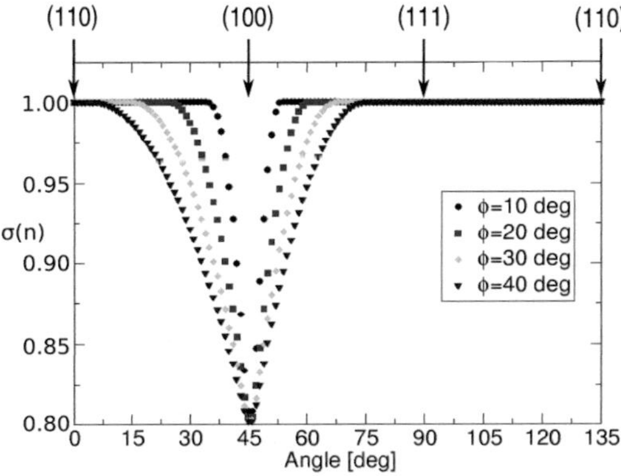

Figure 5.34.: As figure 5.30 for variation of opening angle ϕ in the range of $10°$ to $40°$.

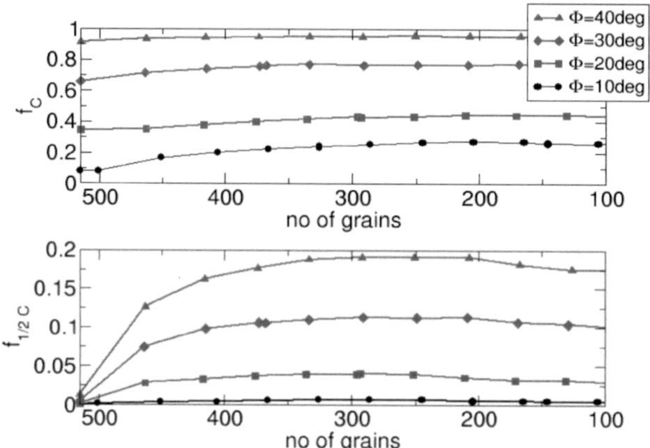

Figure 5.35.: As figure 5.25 for configurations with varied cusp opening angle ϕ.

the energy cusp over the course of the simulation. Again, the fraction of triangles with low energy orientations increases for all investigated configurations during the first third of the simulation time. Afterwards, the fraction stabilizes. Accordingly, the average grain boundary energy $\bar{\gamma}$ decreases for all four configurations during the first third of the simulation period. The average grain boundary energy stabilizes at $\bar{\gamma} = 0.98 \pm 0.03$ for the configuration with a cusp opening angle of $\phi = 10°$, $\bar{\gamma} = 0.96 \pm 0.03$ for $\phi = 20°$, $\bar{\gamma} = 0.95 \pm 0.03$ for $\phi = 30°$ and $\bar{\gamma} = 0.94 \pm 0.03$ for $\phi = 30°$. Neglecting the widest cusp, a linear correlation between the average grain boundary energy and the area fraction of the orientation landscape affected by the cusp in the energy functional can be approximated as

$$\bar{\gamma} = 0.99 - 0.18 \frac{A_C}{A_{tot}}, \tag{5.7}$$

with an error of $\sigma_S = 3.2 \cdot 10^{-3}$ for constant cusp depth of $\Delta\sigma = 0.2$. Here, A_{tot} is the total surface area of the pole sphere and A_C is the part of the surface area affected by the energy cusp.

The relative growth rates, assessed from the normalized area evolution, exhibit accelerated growth for the four configurations with different cusp

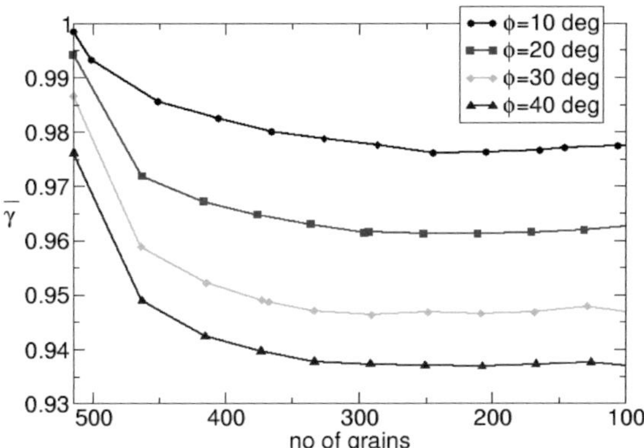

Figure 5.36.: As figure 5.26 for variation of cusp opening angle ϕ.

opening angles compared to the isotropic one. The acceleration increases up to a cusp opening angle of $\phi = 30°$. For wider cusps, the relative growth rate is found to stagnate.

I(d) two-cusp functional The influence of an additional cusp in the applied surface energy functional is studied in two simulation scenarios. The applied functionals differ in the overall attraction zone of the energy cusps:

(i) **extended attraction zone functional:** one cusp functional applied in energy variation scenario (a) with a cusp depth of $\Delta\sigma = 0.2$ and opening angle $\phi = 30°$ provided with additional cusp (cusp depth $\Delta\sigma = 0.15$, opening angle $\phi = 30°$) at the <111> orientation increasing the overall attraction zone.

(ii) **equal attraction zone functional:** surface energy functional as for scenario (i) but with reduced opening angle ϕ of the <100> cusp resulting in an overall attraction zone equal to the one cusp functional applied in scenario I(a).

Stereographic projections of both energy landscapes are given in figures 5.23(b) and (c). Figure 5.37 shows the relative surface energies around

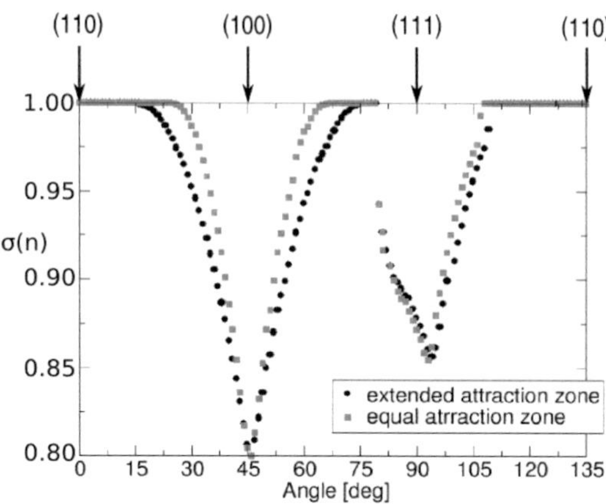

Figure 5.37.: Normal dependent relative surface energies around the perimeter of the standard stereographic triangle for the two-cusp functionals with extended and equal attraction zone.

the perimeter of the standard stereographic triangle. The simulations are started from configurations containing 296 grains.

For both energy functionals, increasing fractions f_C and $f_{1/2C}$ of low energy triangles are observed during the first half of the simulation time, see figure 5.38. After a transition period both fractions remain constant. The system with the extended attraction zone functional exhibits a 10% higher value of both f_C and $f_{1/2C}$ compared to the system with the two-cusp functional with identical attraction zone. Accordingly, the average relative grain boundary energy $\overline{\gamma}$ decreases in the first half of the simulation and then stabilizes at approximately $\overline{\gamma} = 0.935 \pm 0.03$ for the functional with extended attraction zone and $\overline{\gamma} = 0.945 \pm 0.03$ for the functional with equal attraction zone, see figure 5.39.

Comparing the normalized growth dynamics for both two-cusp functionals the one cusp functional with $\Delta\sigma = 0.2$ and $\phi = 30°$ and the isotropic scenario reveals a significant increase in relative growth rate K for the two-cusp functional with extended attraction zone ($K = 1.60 \pm 0.03$), while the relative growth rate for the simulation of the two-cusp functional with

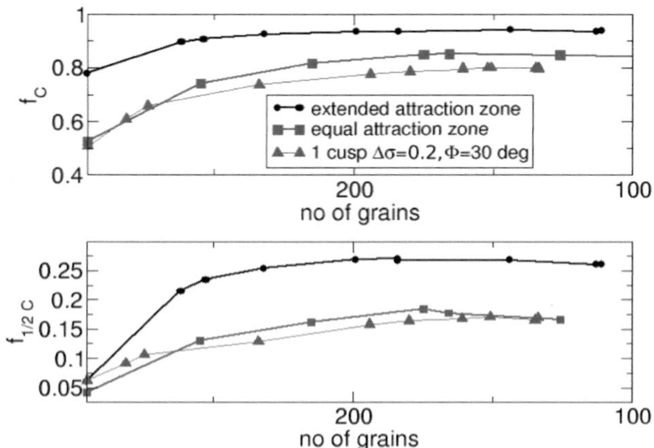

Figure 5.38.: As figure 5.31 for simulation scenarios using two-cusp energy functionals.

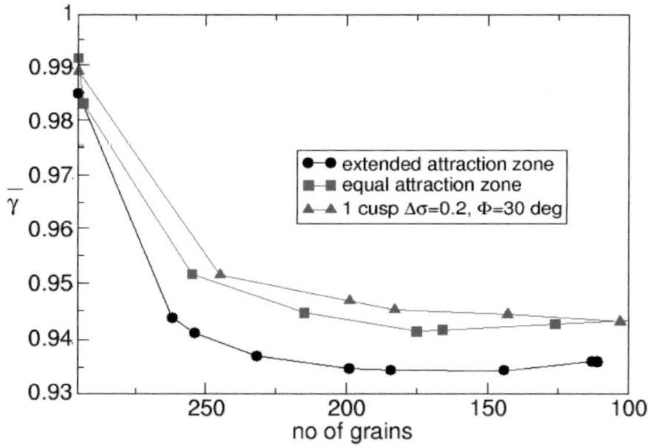

Figure 5.39.: As figure 5.32 for simulation scenarios using two-cusp energy functionals.

equal attraction zone ($K = 1.34 \pm 0.03$) and the relative growth rate K of the one-cusp scenario ($K = 1.38 \pm 0.03$) are equal within the error margins.

II. Mobility Variation

Simulations studying the influence of anisotropic grain boundary mobility on the overall growth dynamics are performed in two scenarios:

> **(a) orientation dependent mobility:** scenario applying an orientation dependent grain boundary mobility functional. Variation of the maximum mobility m_{max}.

> **(b) fraction of highly mobile grain faces:** scenario assigning elevated mobility to varied fractions of entire grain faces.

II(a) orientation dependent mobility Since there is no published data on orientation dependent grain boundary mobility in strontium titanate an artificially designed mobility functional is introduced. For reasons that come into play when studying combined energy and mobility anisotropy (see III. Energy and Mobility Variation), the functional is build such that it is inverse to the applied orientation dependent surface energy functional with opening angle $\phi = 30°$. Relative mobility values around the perimeter of the standard stereographic triangle are given in figure 5.40.

As it is well accepted that the orientation dependent grain boundary mobility may vary by orders of magnitude [9, 18], the maximum normal dependent mobility m_{max} is varied between 1 and 50.

In analogy to the f_C and $f_{1/2C}$ values studied for the energy variation scenarios, the fractions f_P and $f_{1/2P}$ of triangles, which are affected by the mobility peak and the half-height of the mobility peak are investigated. Figure 5.41 shows the f_P and $f_{1/2P}$ values for configurations applying orientation dependent mobility functionals with maximum mobility m_{max} in the range of 1 to 50 plotted against the number of remaining grains. Both values decrease by 4-7% over the course of the simulation regardless of the peak height m_{max}.

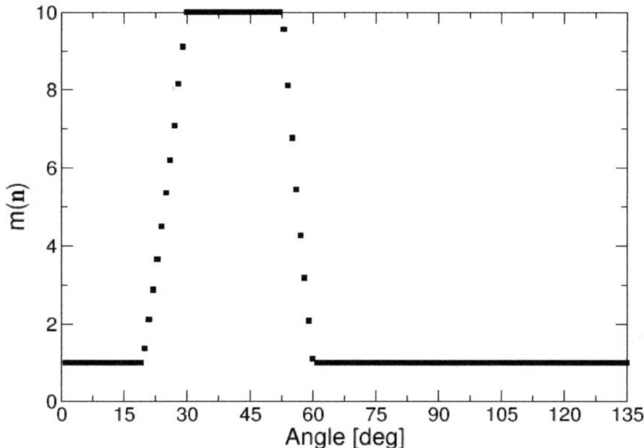

Figure 5.40.: Normal dependent mobility around the perimeter of the unit triangle.

Accordingly, the average grain boundary mobility $\overline{M}$ decreases by 3-8%. The correlation between average grain boundary mobility and maximum surface mobility is

$$\overline{M} = 0.85 + 0.12 m_{max}. \tag{5.8}$$

Figure 5.42 shows the normalized area evolution for systems with orientation dependent mobility in the range of 1 to 50. All configurations show a linear growth behavior. The systems with elevated mobility exhibit accelerated growth. The acceleration of the growth rate is found to slow down with increasing maximum mobility, see figure 5.43.

II(b) fraction of highly mobile grain faces As a prerequisite for the investigation of abnormal grain growth, the influence of varying fractions of high mobility grain boundaries on the overall growth dynamics is investigated. Therefor, fractions f_{GF} of 10% to 90% of the grain faces are provided with a relative surface mobility M=20. The highly mobile grain faces are chosen randomly.

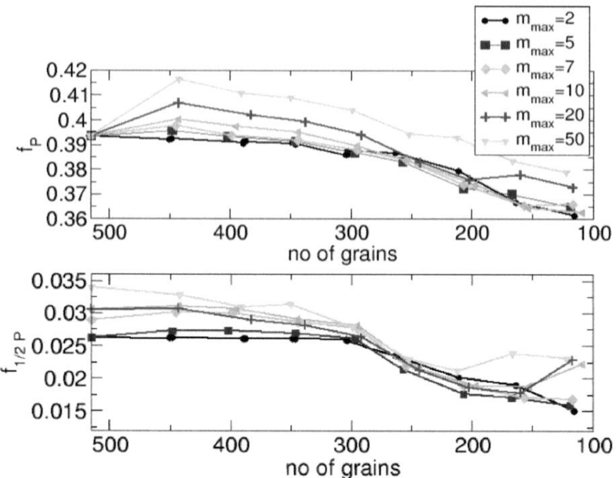

Figure 5.41.: Fractions f_P and $f_{1/2P}$ of triangles affected by the mobility peak and the half-height of the mobility peak respectively plotted against the number of remaining grains.

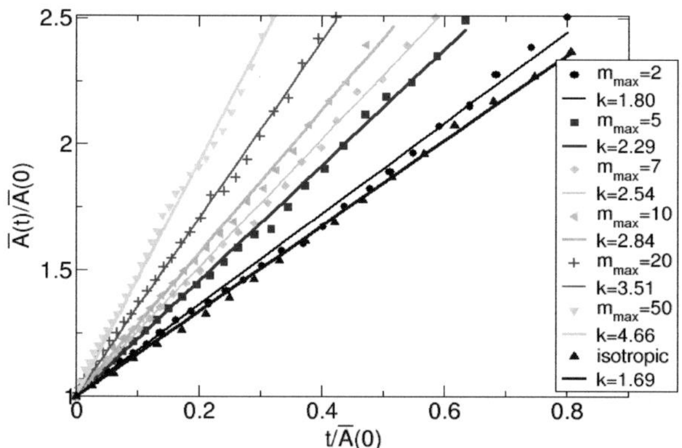

Figure 5.42.: As figure 5.24 for systems with orientation dependent mobility anisotropy under variation of the maximum normal dependent mobility m_{max}.

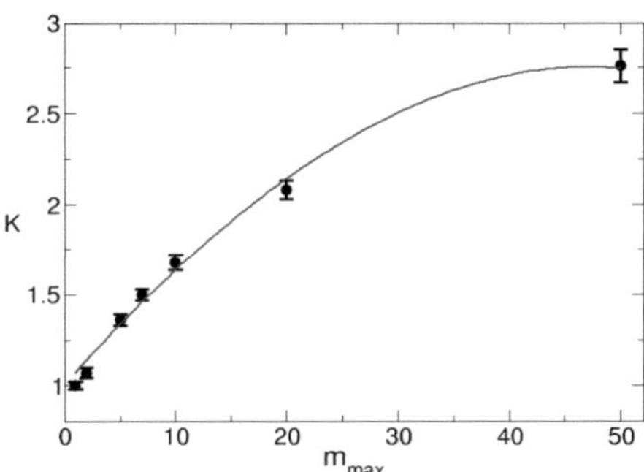

Figure 5.43.: Relative growth rates for systems with orientation dependent mobility and varied peak height m_{max} plotted against maximum orientation dependent mobility m_{max}.

Grain faces generated due to topological changes during grain growth are assigned low ($M{=}1$) or high ($M{=}20$) mobility with a probability corresponding to the overall distribution in order to keep a constant fraction of special boundaries.

For all fractions f_{GF} the normalized grain area evolves linear with time. The relative growth rates K for scenarios with different fractions f_{GF} of highly mobile grain boundaries are shown in figure 5.44. Two separate regions with a transition at $f_{GF} = 0.5$ are identified. Both regions are best approximated by a linear regression. For mobile grain face fractions of up to 0.5 the relative growth rate K increases by a factor of 4.92. For higher fractions of highly mobile grain faces, the slope of the regression line rises to 31.82, finally yielding a twenty-fold increase for the system with high mobility grain faces only.

Figure 5.45 shows normalized grain size distributions for systems with different fractions f_{GF} of highly mobile faces. All distributions are found to remain self-similar throughout the coarsening process. For intermediate fractions the peak is slightly lowered with an overall broadening. For the limiting fractions $f_{GF} = 0$ and $f_{GF} = 1$ the distributions are almost identical. This is expected because effectively these two systems are both treated with isotropic grain boundary mobility.

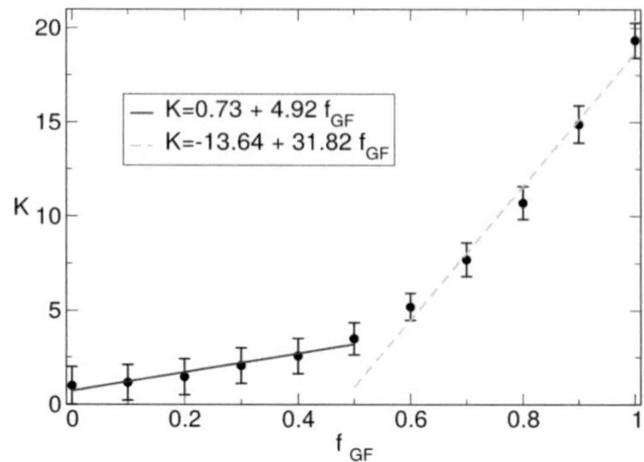

Figure 5.44.: Relative growth rate K plotted against fraction of high mobility grain faces f_{GF}.

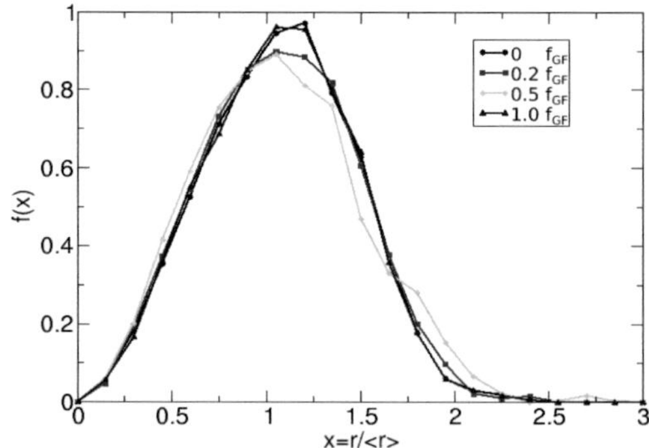

Figure 5.45.: Normalized grain size distribution for different fractions $f_{GF}=0$, 0.2, 0.5, 1 of high mobility grain faces

III. Energy and Mobility Variation

Simulations applying combined orientation dependent mobility and energy functionals are performed in order to compare their influence on the growth dynamics to pure energy and mobility anisotropy effects.

The combined energy and mobility anisotropy is simulated two ways, differing in the assigned mobility. The low energy facets are either provided with an elevated mobility ($m_{max} = 10$) or a lowered mobility ($m_{min} = 0.5$). These combined anisotropy scenarios are compared to the pure energy anisotropy scenario with $\Delta\sigma = 0.2$ and $\phi = 30°$ and the pure mobility anisotropy scenario with $m_{max} = 10$. Figure 5.46 shows the normalized area evolution for all scenarios. Compared to the isotropic scenario, all anisotropic configurations exhibit·accelerated growth rates. The increase in relative growth rate K is most pronounced for scenarios combining reduced grain boundary energy with elevated mobility. At the same time, a combination of reduced grain boundary energy and reduced mobility weakens the increase in growth rate. The relative growth rate for combined energy and elevated mobility ($K = 2.09 \pm 0.03$) is approximately the product of the growth rates for pure energy ($K = 1.36 \pm 0.03$) and pure mobility anisotropy ($K = 1.68 \pm 0.03$).

Distributions of the number of neighbors F_G of all grains are compared between the different anisotropy scenarios and the isotropic scenario in figure 5.47. For the anisotropic systems the distributions are generated after the structure is coarsened to 250 grains.

No significant deviation between the distributions is observed. The average number of neighbors per grain $\overline{F_G}$ is 13.6 for the isotropic scenario, 13.8 for the system treated with pure energy anisotropy and 13.7 for the remaining scenarios.

Distributions of number of edges per grain generated for the same scenarios and simulation steps as for the number of neighbor investigation are presented in figure 5.48. There is no significant deviation between the edge distributions for the anisotropic and isotropic scenarios.

The average number of edges per grain $\overline{E_G}$ is 34.8 for the isotropic configuration, 35.4 for pure energy as well as combined energy and mobility

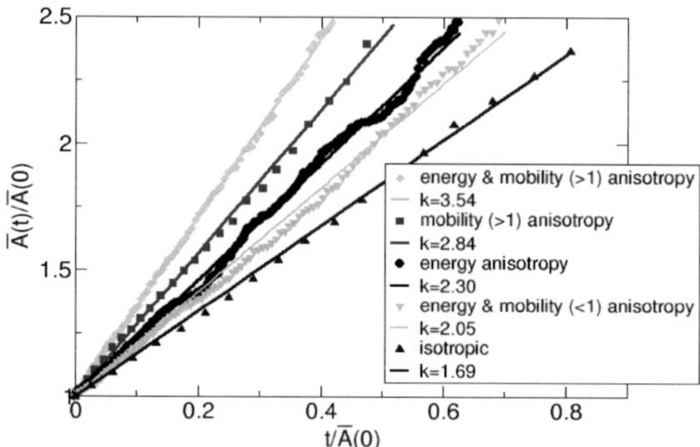

Figure 5.46.: As figure 5.24 for systems with energy, mobility and combined energy and mobility anisotropy.

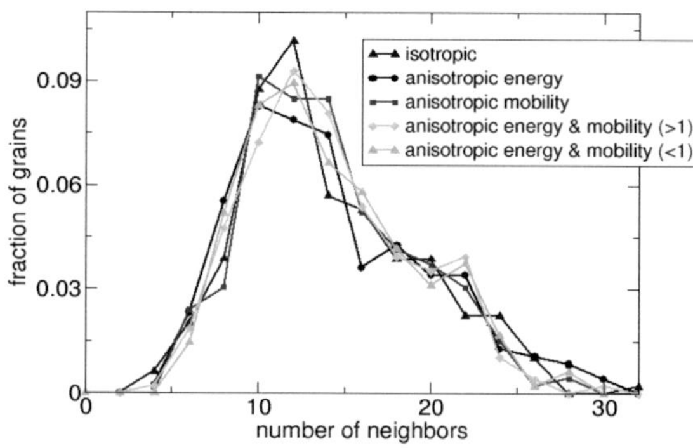

Figure 5.47.: Distributions of number of neighbors F_G for different grain boundary property anisotropy scenarios.

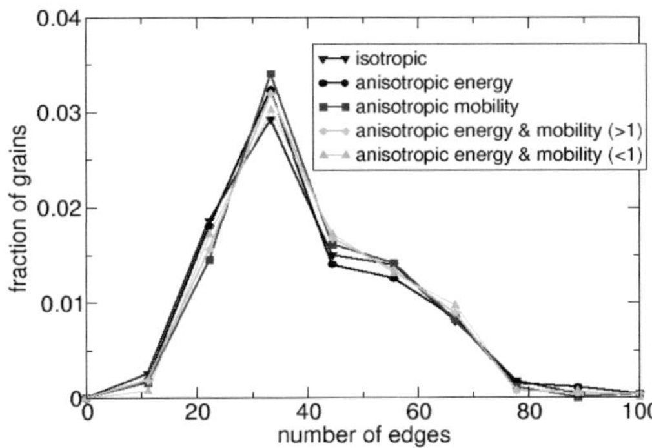

Figure 5.48.: Distributions of number of edges for different grain boundary property anisotropy scenarios.

anisotropy with reduced mobility, 35.2 for pure mobility anisotropy and 35.1 for combined energy and mobility anisotropy with elevated mobility.

The evaluation of the average relative grain size $\overline{g}_C$ related to the number of neighbors per grain given in figure 5.49 shows a linear correlation for the energy anisotropy scenario. All other scenarios can not be regressed linearly.

The influence of the anisotropy scenarios on the grain shape is investigated comparing sphericity distributions, see figure 5.50. Compared to the distribution for the isotropic scenario, a slight broadening of the distribution around the same mean value ($\overline{\Psi} = 0.87$) is obtained for all anisotropic scenarios, except for the pure energy anisotropy cases accounting for torque. Here, the broadening of the distribution is more distinctive and the mean value is shifted towards lower values ($\overline{\Psi} = 0.81 - 0.83$). Both, the shift and the broadening are found to be more distinct with increasing cusp depth $\Delta\sigma$.

Figure 5.51 shows the average sphericity $\overline{\Psi}$ over the course of the simulation for all six configurations. The average sphericity stays almost constant during the whole time span of the simulation for all configurations, except for the one receiving pure energy anisotropy under consideration of torques. This configuration exhibits a significant decrease in average sphericity. A distinct difference in grain shape is also visible in 2D cross-sections of

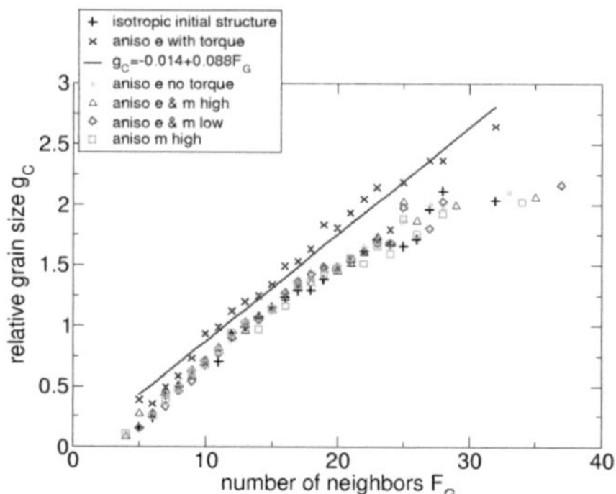

Figure 5.49.: Relative average grain size $\overline{g}_C$ plotted against number of neighbors F_G for the initial structure and anisotropic simulations with and without torque contribution after coarsening to 250 grains. The continuous line is a linear regression to the data points obtained in the energy anisotropy scenario accounting for torque contributions.

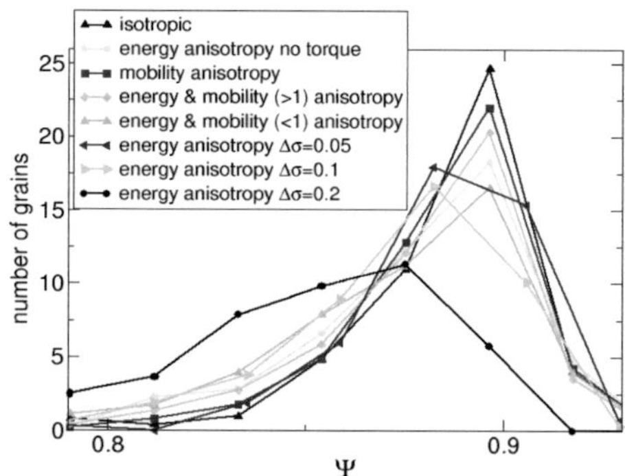

Figure 5.50.: Number count of sphericity values after coarsening to half the initial number of grains for simulation scenarios with energy anisotropy with and without torque contribution, mobility and combined energy and mobility anisotropy compared to the isotropic case.

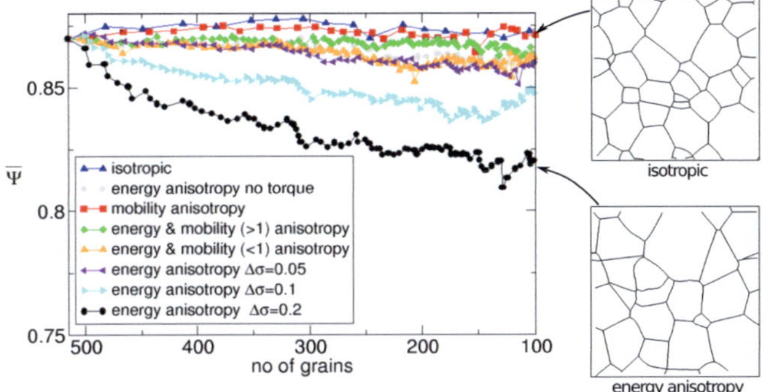

Figure 5.51.: Average sphericity $\overline{\Psi}$ plotted against number of remaining grains for simulation scenarios with energy anisotropy with and without torque contribution, mobility and combined energy and mobility anisotropy compared to isotropic scenario. The inlays show cross sections of the different structures (isotropic and anisotropic grain boundary energy).

the simulated structures, given as inlay in figure 5.51. Compared to the isotropic structure (upper image), the structure coarsened with orientation dependent grain boundary energy under consideration of torques reveals straighter boundaries with dihedral angles deviating severely from 120°.

5.2.2. Abnormal Grain Growth

Conditions for the nucleation of abnormal growth under combined energy and mobility advantages are investigated providing candidate grains of varying initial size with relative grain boundary energy and mobility values in the range of $\frac{\gamma}{\overline{\gamma}} = 0.8 \ldots 1.2$ and $\frac{M}{\overline{M}} = 1 \ldots 20$. Each class of grains is represented by a randomly chosen candidate grain. The remaining grains (matrix grains) are treated isotropically. The size evolution of the candidate grains is compared to the average trend. A candidate grain is considered as growing abnormal, when the grain exhibits a relative growth rate

$$\frac{d(r/\overline{r})}{dt} > 0 \tag{5.9}$$

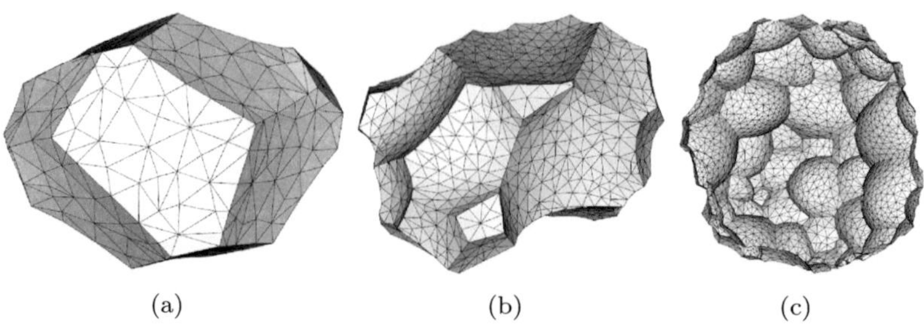

Figure 5.52.: Candidate grain treated with a relative grain boundary energy of 0.8 and a relative grain boundary mobility of 10 at different steps during its evolution. The grain has been rescaled. It consists of (a) 14, (b) 39 and (c) 158 faces respectively.

and its normalized grain size $\frac{r}{\bar{r}}$ exceeds a factor of three. An example of a candidate grain with 14 grain faces at three stages during the simulation is shown in figure 5.52. The particular grain is treated with a relative grain boundary energy of 0.8 and a relative grain boundary mobility of 10. Upon visual inspection the grain develops the characteristic abnormal grain shape with an overall convex but locally concave shape resulting from the smaller neighboring grains. Figure 5.53 shows the probability p for the occurrence of abnormal growth as a function of relative grain size and mobility for relative energy values $\frac{\gamma}{\bar{\gamma}} = 0.8$ and $\frac{\gamma}{\bar{\gamma}} = 1.2$. Red data points mark the probability assessed as percentage of abnormally growing candidate grains of a particular relative size. Grey data points mark interpolations for probability values 0.1, 0.5 and 0.9, respectively. Additionally, the stability range of abnormal growth from the well-accepted mean field approach [36] is given as continuous blue line.

For both scenarios, the transition from the region of normal to abnormal growth is rather gradual, the spread in $\frac{r}{\bar{r}}$ between the probability of 10, 50 and 90% for abnormal growth does not show a systematic dependency on the relative grain boundary properties $\frac{\gamma}{\bar{\gamma}}$ and $\frac{M}{\bar{M}}$. The transition range, where the probability p increases from 10% to 90%, is about 0.6 wide on the $\frac{r}{\bar{r}}$ axis. For $\frac{\gamma}{\bar{\gamma}}=0.8$ the mean field line is between 50 and 90% probability, for $\frac{\gamma}{\bar{\gamma}} = 1.2$ the line is close to the 10% value.

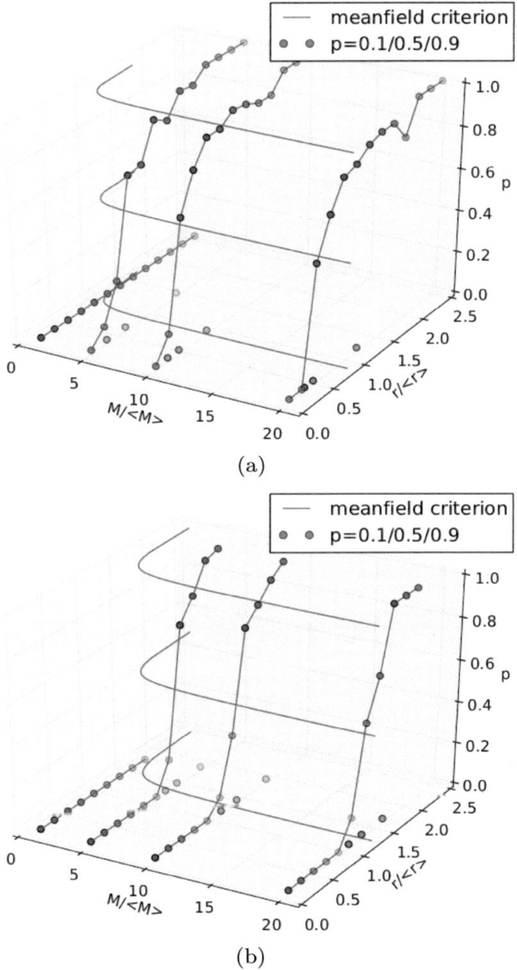

Figure 5.53.: Probability p for the occurrence of abnormal grain growth as a function of relative grain boundary mobility $\frac{M}{\langle M \rangle}$ and relative grain size $\frac{r}{\langle r \rangle}$ of candidate grains: with relative grain boundary energy (a) $\frac{\gamma}{\langle \gamma \rangle} = 0.8$ and (b) $\frac{\gamma}{\langle \gamma \rangle} = 1.2$. Solid lines indicate the stability range of abnormal grains in mean field theory [36] plotted at different p values for guiding the eye. The points in the $p = 0$ plane indicate the projected probability (0.1, 0.5 and 0.9) of the simulation results, for better comparison with the mean field plot.

6. Discussion

6.1. Microstructure Characterization

In the context of the present work, the grain, which is described by its size, shape and crystallographic orientation, is the fundamental microstructure element. The discussion of microstructure characterization is organized accordingly.

6.1.1. Grain Size and Morphology

Since the grain size distribution is a sensitive metric of polycrystalline structure great importance is attached to the reliability of grain size measurement. Prior to adressing more complex correlations, the accurateness and comparability of the grain size measurements in the various investigations presented in this thesis must be critically reviewed.

In section 5.1.1 the average grain size of the sample material in the initial state was assessed as average grain radius $\bar{r}$ by measuring the equivalent circle diameter on micrographs of the bulk material as well as cross sections of the DCT reconstruction. Both values are equal within the error margins of $\pm 2\,\mu\text{m}$.

For the DCT reconstruction, the average initial grain size was also derived from the volume information, approximating the grains as spheres. The factor between the values obtained by the 3D and 2D method can be explained with stereological effects arising from cutting 3D polyhedra with apparent diameter smaller than the maximal diameter. With a value of $\frac{\bar{r}_{2D}}{\bar{r}_{3D}} = 0.9$ the ratio of equivalent circle diameter and equivalent sphere diameter measurements differs by about 10% from the values found in literature (0.78 to 0.84) [95, 96]. However, the difference between the 2D and 3D value for the average grain size in the DCT reconstruction of the initial structure is small compared to the possible maximal error introduced by the dilation step of the DCT reconstruction. Here, an average radial dilation of 2.6µm per grain is applied in order to reach a dense material.

In the course of a validation of the DCT reconstruction of the annealed state against EBSD data, the grain size and shape of two corresponding EBSD and DCT sections were compared in figure 5.17. The number of grains found in the EBSD and corresponding DCT maps is almost equal, the average grain size assessed by the linear intersect method is equal within the error margins. The difference in the average grain size is the result of a number of grains that are barely cut in either of the cross sections thus appearing as very small areas. Furthermore, it is more likely to find less grains in the DCT maps since diffraction spots with an area smaller than about 100 pixels have been excluded from the current analysis for the purpose of noise reduction. As a consequence grains below the corresponding volume (10^3 voxels) can not be indexed and are lost in the analysis. The smallest detected grain has a radius of about 2.2µm in the undilated and 2.7µm in the dilated state. This gives a realistic estimation for the resolution limit for the reconstruction of small grains. Also, already a small angular uncertainty in the identification of the best fitting plane in the DCT reconstruction might leave very small grains undetected above or below the cutting plane.

Finding a good agreement in overall grain size, the average euclidean distance between the grain boundary networks obtained by DCT and EBSD was found to be 1.95µm and 1.98µm, see section 5.1.2. These values correspond well to the previously reported accuracy of 1.60µm obtained from a 3D distance transform [74]. It shall be emphasized that an estimate made from a 2D distance transform on a 3D grain boundary network is necessarily conservative. For grain boundaries which are close to parallel to the 2D cross section, a one pixel error in the 3D reconstruction may give rise to large shifts and irregular shapes in the corresponding 2D observation. The brown grain at the bottom of figures 5.17(a) and (b), as well as the light green grain in the center of figure 5.19(d) are examples for this kind of configuration. Close ups of these two regions as obtained from the fitted DCT cross sections with a separation distance of 3µm are given in figure 6.1. The remaining differences in grain boundary location, which tend to appear half way between two triple points, can partly be

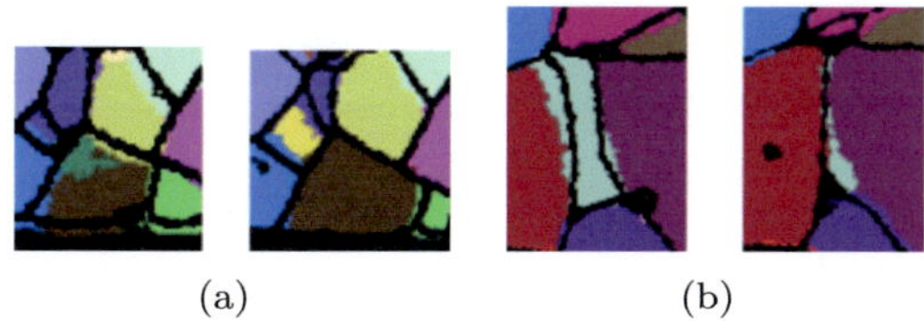

Figure 6.1.: Close ups of two grains in two cross sections of the DCT reconstruction. Both the brown grain in the lower part of (a) and the green grain in the center of (b) exhibit boundaries which are close to parallel to the section resulting in a large shift and irregular shape when going from one section to the other.

explained by the dilation procedure applied during post processing of the diffraction data.

This dilation, being of uniform character, hinders the detection of pronounced faceting expected in anisotropic materials. Such faceting is detected in EBSD sections. A close up of a faceted grain boundary that is curved in the reconstruction can be seen in the purple grain in figure 5.19(d). Another fact, that might account for some of the deviations in the grain boundary networks is the manual preparation of the specimen resulting in an uneven cutting plane. The one voxel wide extrusion-like artifacts obtained at some grain boundaries in the DCT grain map, see for example figure 5.19(b) and (d) can be traced to the interpolation procedure applied during plane fitting.

Porosity data for the bulk material in the initial state was gained by the buoyancy method as well as from absorption contrast tomography measurements on the DCT specimen. The obtained volume percentages of porosity differ by 15%. The lower amount of porosity obtained from the reconstructed microstructure can be explained by intragranular pores that could not be resolved using absorption tomography but are visible in SEM images of the same sintering load, as well as EBSD sections of the annealed structure, see figures 5.5(b) and 5.17(a) and (b). Only the phase contrast tomography (PCT) investigations applied to the annealed state of the specimen were able to resolve these smaller pores inside the grains.

Comparing the pores obtained in EBSD investigations of the annealed state and corresponding sections of the microstructure reconstruction reveals a mixed picture: while notably even small intragranular pores were observed in the PCT reconstruction, some pores appear to be at doubtable location inside the reconstructed microstructure and inconsistently connected to the grain boundaries, see close ups and textual description of figure 5.19.

A comparison of euclidean distance transforms of the corresponding sections with and without pores is given in figure 6.2. It shows clearly, that despite the higher resolution resulting from PCT data, the porosity obtained by the two characterization methods is not yet reliably comparable. The deviation between the compared grain boundary networks can be traced to several factors: The higher amount of (especially intragranular) pores in the EBSD slices can be related to ring artifacts and uncertainties in the PCT reconstruction [97]. These artifacts hamper the segmentation of individual pores. During pore segmentation a threshold was applied that eliminated small pores and pores with a low intensity value. Furthermore, thin material layers above pores might have collapsed during the mechanical preparation for sectioning and hereby revealed underlying pores, so that the EBSD sections might also contain pores, which are underneath the actual section. Moreover, material that has been removed during polishing might accumulate in large pores, altering the shape of the pores visible in the EBSD orientation map.

Despite the above mentioned elements of uncertainty, the EBSD and DCT grain boundary networks correspond well showing an average euclidean distance below 2μm. The maximum euclidean distance of 7μm was obtained midway between two triple points on grain boundaries that appear to be curved in the DCT reconstruction and straight in the EBSD map and is most certainly caused by the grain dilation during volume reconstruction. Accordingly we can assume that the DCT technique is able to determine grain boundaries with a spatial resolution of ±2μm.

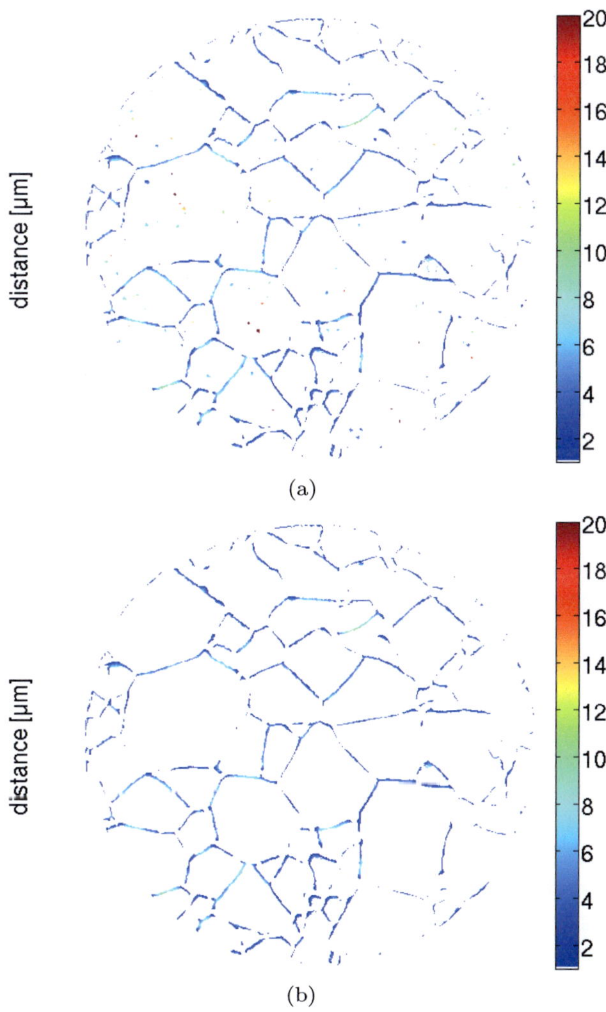

Figure 6.2.: Euclidean distance map as shown in figure 5.18 (a) with and (b) without consideration of intragranular pores. The maximum euclidean distance is significantly reduced in (b). The mean euclidean distance is only changed a little (as shown in section 5.1.2).

6.1.2. Topology

The grain topology in dense polycrystalline microstructures depends on the number of contiguous neighbors, the form of the shared faces between the grains and how these faces are connected at the triple lines. The topology and its evolution is governed by the local equilibrium conditions at the triple lines on one hand and the requirement to build a space filling structure on the other hand.

From the DCT microstructure reconstruction it is possible to investigate the topology of the bulk material and its evolution in 3D. The average number of neighboring grains $\overline{F}_G$ was found to be approximately 13 in both annealing states. The average number of edges per grain $\overline{E}_G$ was found to be 31.1 in the initial and 32.2 in the annealed state. The distribution of the number of edges per grain is best approximated by a log normal fit, see figure 5.14. Its rather long tail can be explained by the few extraordinary large grains ($F_G > 30$) contained in the structure. A comparison of the topological quantities with the values obtained from vertex dynamics simulations ($\overline{F}_G$=13.6-13.8, $\overline{E}_G$=35.1-35.4, depending on simulation scenario) and statistical grain models such as Coxeter polytopes [98] ($\overline{F}_G$=13.7, $\overline{E}_G$=34.9), Kelvin's α-tetrakaidecahedrons [99] or Williams' β-tetrakaidecahedrons [100] (both $\overline{F}_G$=14.0, $\overline{E}_G$=35.7) shows similar values for the average number of grain faces $\overline{F}_G$, while the average number of edges per grain $\overline{E}_G$ is small with respect to all other investigations. This effect can be traced to pancake shaped pores between two grains contributing with two faces to the face counting but only with one edge to the edge counting. A total of 191 of these pores is contained in the reconstruction of the annealed state. Removing these pores from the counting increases the number of edges per grain to $\overline{E}_G$=34.4. Accordingly, both, the average number of faces per grain $\overline{F}_G = 13$ as well as the average number of edges per grain $\overline{E}_G$=34.4 of the investigated microstructure were found to be smaller than the values predicted by statistical grain models. This is in good agreement with values observed in common metals and alloys [101, 102].

6.1.3. Interface Orientations

Being a sensitive metric of polycrystalline structure, the grain boundary character distribution in polycrystals, represented by relative areas of grain boundaries distinguished by their lattice misorientation and grain boundary plane orientations has been studied extensively both in experiment and simulation.

Prior to investigating grain boundary plane orientation in DCT data, the spatial resolution at the grain boundaries must be addressed: Due to the uncertainty induced by the dilation procedure, the orientation of the interface normal of a grain boundary with an average length of $\bar{r} = 14.3\mu m$ can only be estimated within $\pm 10°$. Since we find a good agreement with EBSD grain boundary networks and intergranular pores are consistently connected to the outgoing grain boundaries, see figure 5.5(b), we conclude however, that the true interface normal uncertainty is significantly below $\pm 10°$.

A preference for {100} grain boundary planes was found in microstructure reconstructions of both annealing states, see figure 5.8. These planes were identified as minimum energy planes both in calculations [103, 104] and experiments [82]. Together, this is in good agreement with recent studies reporting, that the lowest energy interfaces dominate the distribution of grain boundary planes in $SrTiO_3$ [3, 105]. The microstructure reconstruction of the initial state reveals grains with a global outer shape close to a <001> oriented cube, see figure 5.7. While the global interface normal orientations in these cubic grains are along <001> directions with respect to the grain reference frame, these faces are composed of many grain boundaries separated by triple lines. A similar behavior has also been observed for grain boundaries surrounding abnormally growing grains in strontium titanate [106]. There, the general interface orientation is parallel to <100> orientations of the abnormally growing grain. These transmission electron microscopy observations indicate that the grain boundaries are curved close to triple lines. Boundary segments away from these triple lines can be atomically flat over micrometers [106, 107]. In the current study no such well developed abnormal grains were present.

The <100> orientation was also observed to be the preferred orientation for external crystal surfaces in $SrTiO_3$ [82]. This justifies to derive the

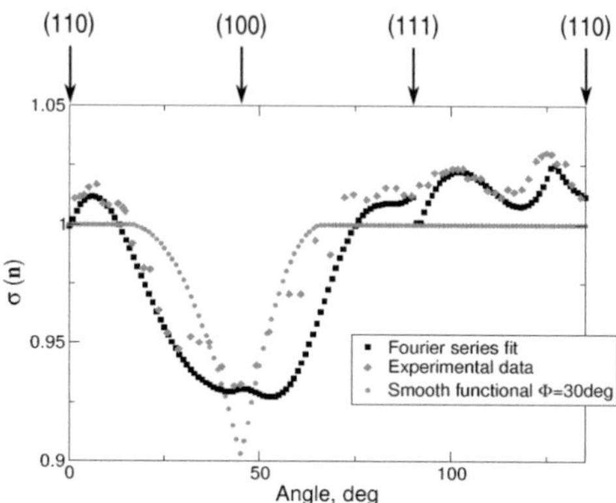

Figure 6.3.: Experimentally obtained and Fourier fitted relative surface energies around the perimeter of the standard stereographic triangle compared to the smooth energy functional with cusp opening angle $\phi = 30°$.

orientation dependent interface energy from the surface energy functional presented in [82]. A smooth version of this functional was used to estimate the orientation dependent interface energy for the vertex simulations. As can be seen in figure 6.3, the smooth functional with cusp opening angle 30° mimics the overall form of the deepest and steepest energy cusp better than the Fourier fit proposed in [82]. We assume the most relevant influence on microstructure evolution to be associated with this <100> cusp.

There is no consensus in literature about the misorientation distribution in polycrystalline $SrTiO_3$. A preferred occurrence for $\Sigma3$ boundaries reported in [108] is in contrast to a random misorientaton distribution obtained in [3]. The misorientation distribution obtained from DCT experiments is random, changes in the distribution of misorientation angles during annealing were below the resolution, see figures 5.12(a) and (b). This is expected for undeformed polycrystals with rather large grain size.

Overall, these findings enforce the view that the grain boundary population in polycrystalline ceramics is correlated with the grain boundary plane orientation rather than the lattice misorientation as previously reported [3, 16].

6.2. Microstructure Evolution

The time evolution of the grain size during annealing of the tomography specimen analyzed in terms of a quadratic growth law

$$k = \frac{A(t) - A(t_0)}{\Delta t} \tag{6.1}$$

gives $k = 2.3\ 10^{-14}$. Annealing experiments with identical heating conditions using scanning electron microscopy for the microstructure characterization of 386 grains [84] report a value of $k = 5.6\ 10^{-14}$. The deviation of more than 40% can be explained by the uncertainty in the grain size measurement ($\pm 2\mu m$) accounting for an error of 14%. The reduced growth rate obtained for the tomography specimen might also be rooted in a pinning effect due to surface grooving that reaches deeper than the one layer of surface grains that have already been excluded from the analysis. Due to the limited reproducibility of the heating conditions (we used two different furnaces in different laboratories), the tomography specimen might have been annealed at slightly lower temperatures. Moreover, we have to consider the small feature size in both experiments.

The volumetric grain size distributions obtained at both annealing states shown in figure 5.2 reveal a significant broadening and a shift towards smaller relative grain sizes. This is in agreement with the grain size distribution evolution obtained during annealing experiments on dense $SrTiO_3$ specimens that have been sintered at 1450°C in oxygen atmosphere [2]. In [2] a significant shift of $0.25\frac{r}{\bar{r}}$ (approximately one binning width) of the maximum towards smaller relative grain sizes was observed over the course of the first 1.5 hours during annealing at 1550°C. Considering the growth behavior of $SrTiO_3$, which does not follow a single Arrhenius dependency over the whole range of annealing temperatures, it is not straightforward to compare the annealing behavior of specimens annealed at different temperatures. In the present case however the initial microstructures were both sintered and annealed at temperatures belonging to the high temperature growth regime, see left slope in figure 3.7, so that a comparison of these experimental results is legitimate. In order to assess the observed shift of the peak quantitatively a distribution function needs to be applied. The equivalent circle diameter grain size distributions of both experiments are best approximated by log normal fits. A comparison of these fits is given in

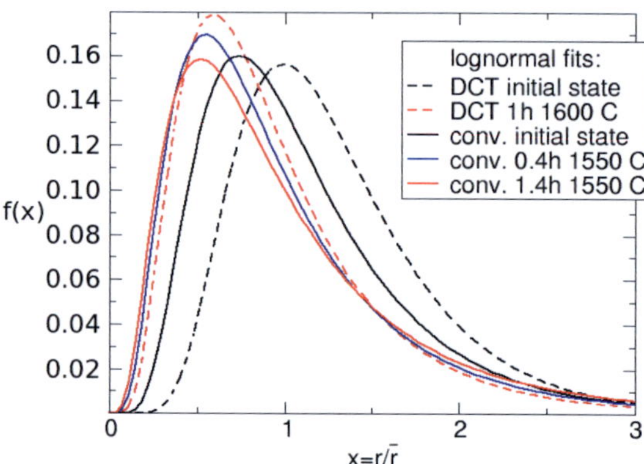

Figure 6.4.: Log normal fits to grain size distributions of 2D cross sections during annealing. Dashed lines: grain sizes obtained by DCT before and after ex-situ annealing at 1600°C. Solid lines: grain sizes at three stages during annealing at 1550°C as obtained by optical microscopy [2].

figure 6.4. Both, the broadening and the shift of the grain size distribution during annealing were also reported in experimental investigations of the coarsening process in the final sintering stage of alumina [109] and during annealing of $BaTiO_3$ [110].

Both effects, a broadening and a maximum shift of the distribution towards smaller relative grain sizes are also observed in grain growth simulations with orientation dependent interface energy: Figure 5.29 shows the grain size distribution of the isotropically coarsened initial structure as well as the distributions after coarsening to 250 remaining grains for simulations with anisotropic grain boundary energy with and without torque contribution. Both anisotropic configurations exhibit a broadening and a maximum shift of the distribution towards smaller relative grain sizes. The broadening is more pronounced for the simulation accounting for torque contribution. This behavior is contrary to results obtained by 2D phase field modeling [111] and simulations using a 3D Monte Carlo Potts model [112]. A possible origin of the discrepancy between those two investigations and the ones presented in this work is the handling of energy anisotropy in the simulations. The phase field as well as the Monte

Carlo Potts Model do not explicitly take into account torque contributions due to inclination dependent grain boundary energy.

Simulations applying orientation dependent grain boundary energy showed, that both grain size and shape may differ from the values observed in the isotropic configuration. One example for a quantity that is strongly affected by the energy anisotropy is the equilibrium angle distribution given in figure 5.27. A broadening in the distribution is observed for all anisotropic configurations. This is expected from the latter part of the Herring relation 2.9 and in good agreement with simulations applying a Monte Carlo model [113]. The more pronounced broadening for the simulation scenario accounting for torque contribution can be explained as resulting from a torque-driven rotation of boundaries. Instead of building the 120° equilibrium angle, the boundary segment connected to the triple line is exposed to a momentum trying to rotate it into an energetically more favorable position. This is also true for dicretizational units not shared by a triple line and reflected in the increasing fraction of low energy orientations that was observed for scenarios including torque contributions exclusively, see figure 5.25.

A linear correlation between relative average grain size and number of neighbors is found for the reconstructed microstructure in both annealing states, see figure 5.15. The slope of the regression lines of both states does not vary more than 10%. This is in good agreement with findings in other sintered materials [114, 115] and gives rise to the suggestion, that the grain shape distribution remains self-similar throughout annealing. A linear correlation is also observed in the vertex dynamics simulations featuring pure energy anisotropy under consideration of torque contributions. All other scenarios, independent of the kind of anisotropy included, were found to show a declining rise in relative average grain size $\bar{g}_C$ with increasing number of neighbors F_G, see figure 5.49. This is expected for structures with predominantly spherical grains where the average grain size is proportional to the grains volume and the number of faces is proportional to the grains surface area. Accordingly the shape of the grains in the energy anisotropy scenario is expected to diverge from the sphere shape. Grains within one class are relatively bigger, which intends a more convex grain shape compared to isotropic grains of the same class. The slope of the regression line for the energy anisotropy scenario was found to remain constant once the system is transitioned to the anisotropic state.

Considering this and the fact that the simulations accounting for torque contributions must be regarded as closest to the real physical situation, the grain shape in anisotropic structures can be assumed to vary from that in isotropic ones particularly for grains with a high number of faces. This is confirmed by sphericity values obtained from the simulation data for all scenarios, showing a reduced average sphericity $\overline{\Psi}$ and hence a more cubic grain shape for structures simulated with pure energy anisotropy, see figures 5.50 and 5.51. Here. the change in grain shape is also visible from the inlays showing 2D cross-sections of the simulated structures. The insertion of the anisotropy to an isotropically primed initial structure causes the grain shape to evolve from a spherical shape towards a more cube-like shape. The average sphericity obtained in the simulation scenario dealing with anisotropic grain boundary energy under consideration of torques reduces from 0.87 to 0.82 (a cube corresponds to $\Psi = 0.8$). This is in good agreement with the average sphericity value of 0.81 observed in both annealing states of the DCT experiments and the sphericity distribution given in figure 5.16.

It is not straightforward, that structures simulated with combined energy and mobility anisotropy come to resemble isotropic ones. A possible hint towards an explanation is found in the comparison of fractions of in-cusp orientations for the different scenarios given in figure 6.5. While the fraction of in-cusp orientations increases throughout the simulation for the scenario with pure energy anisotropy, the fraction stays roughly constant, when low energy orientations are provided with reduced or elevated mobility. For the reduced mobility case, it can be assumed that the reduced mobility hinders the motion of grain boundary segments attracted by the low energy orientation thus hindering a change in the grain boundary character distribution. Accordingly, the structure of systems with reduced mobility of low energy grain faces is closer to isotropic ones than that of the scenario with pure energy anisotropy. In the case of elevated mobility, a slight decrease of in-cusp orientations is observed. At the same time, for this scenario, the fraction of grain faces that is attracted by the half-depth width of the cusp approaches zero. Apparently, the extinction of grain faces due to high mobility which was also observed in all simulation scenarios with pure mobility advantages, see 5.41, outrules the torque effect. This phenomenon might be reinforced by the different shape of the energy and mobility peak: the mobility peak has the same attraction zone, but a very steep slope, see figure 5.40. Accordingly all grain faces affected by the energy cusp receive the maximum mobility. Due to the absence of experimentally observed grain boundary mobility functionals in literature, shape and height of the mobility peak were chosen without much physical reasoning. Both aspects should certainly be adressed in future investigations.

Figure 6.6 shows cross sections through the structures simulated with the different anisotropy scenarios at three stages throughout the simulation. The cross sections are colored according to the local grain boundary orientation. Both scenarios including mobility anisotropy exhibit fewer low-energy faces. Moreover, there are a couple of effects which are only visible for the pure energy anisotropy scenario: Apparently, only the pure energy anisotropy scenario allows for grain faces to bend themselves towards a favorable orientation, as can be seen in the grain faces marked by the letter A in figure 6.6. While a gradient in color is visible in the pure energy scenario, all grain faces in the other two scenarios seem to be uniformly colored. This effect is also visible in the grain face below the letter C: while the grain faces stys straight ind all evolution states of both energy and mobility anisotropy scenarios, it is clearly curved in the cross section taken at 290 remaining grains in the pure energy scenario. Once the grain faces are oriented energetically favorable, they appear to be very stable in the pure energy scenario, see grain faces around the letter B. Lastly, we observe more triple line angles deviating from 120° in the pure energy scenario, see shape of the grain encircling the letter C in figure 6.6. This is effect is confirmed by equilibrium angle distributions which are found to be more narrow for simulations with combined energy and mobility anisotropy, see figure 6.7.

These observations do not explain the fact that an additional mobility anisotropy deletes the effects of energy anisotropy in the simulation scenarios. However, it can be followed that a detailed investigation of triple line forces and cusp shape variation is needed to study this phenomenon in greater detail.

The fraction of $\{100\}$ grain boundary planes was found to increase during 1h annealing at 1600°C, see figure 5.8. This is in excellent agreement with 3D simulations, where the fraction of $\{100\}$ grain boundary planes increased for scenarios exhibiting an energy minimum at this orientation,

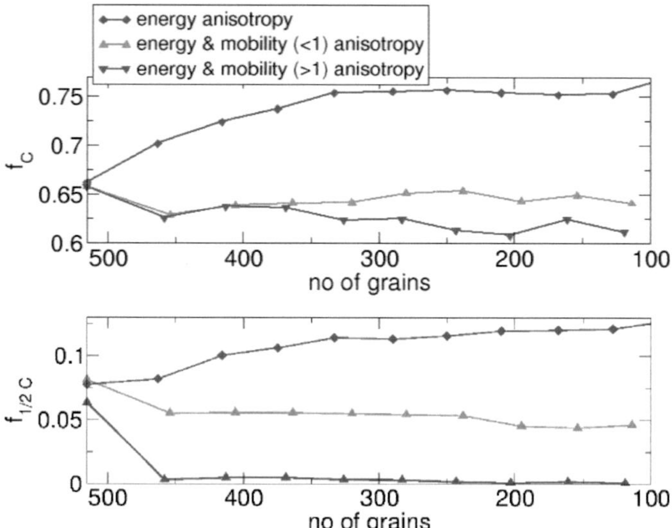

Figure 6.5.: Fractions f_C and $f_{1/2C}$ of discretizational units with grain boundary normal falling in the attraction zone and the half-depth attraction zone of the energy cusp plotted against number of grains remaining in the system for simulation scenarios with combined energy and mobility anisotropy.

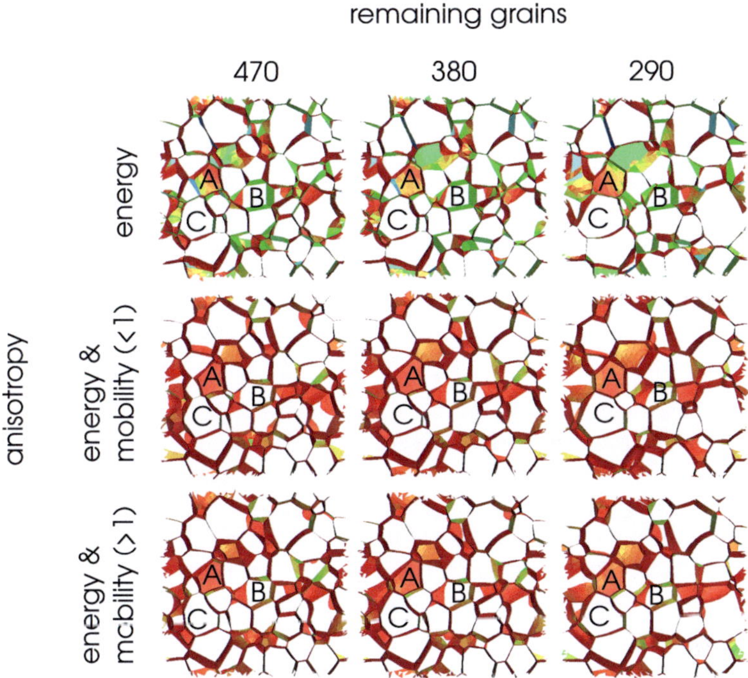

Figure 6.6.: Cross sections of structures simulated with pure energy and combined energy and mobility anisotropy at three stages during grain coarsening. The cross sections are colored according to local grain boundary energy.

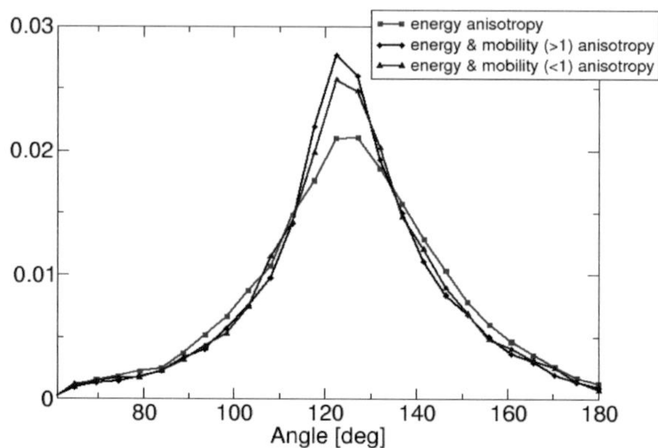

Figure 6.7.: Normalized equilibrium angle distribution for simulations with pure energy and combined energy and mobility anisotropy.

see figure 5.25. As discussed above, this effect is only observed, when torque contributions, helping to rotate grain faces into the low energy orientation, are considered. Varying the depth $\Delta\sigma$ of the cusp in the energy functional for a constant cusp opening angle $\phi = 30°$ shows an increasing fraction f_C of in-cusp orientations, see figure 6.8. The dashed line represents the natural lower limit of f_C given by the fraction of a random orientation distribution falling in the part of the pole sphere affected by the energy cusp with opening angle $\phi = 30°$. Sections through the corresponding grain structures after coarsening from 515 to 250 grains visualize the change in structure affected with the different fractions f_C of low energy orientations. The grain boundary network is colored according to local grain boundary energy. For cusps deeper than $\Delta\sigma = 0.15$, it seems to saturate. At this point 70% of the grain faces are oriented in directions representing only 25% of the orientation space. However, in order to have a consistently connected grain boundary network some non-cusp orientations must be included in the network, too. Clearly, an increasing occurrence of straight grain boundaries and a divergence of the dihedral angles from $120°$ is obtained with increasing fractions f_C of low-energy orientations.

Higher fractions f_C of in-cusp orientations can be reached when varying the opening angle ϕ and thus the fraction $\frac{A_C}{A_{tot}}$ of the orientation space affected

$\phi = 30°$	$\Delta\sigma = 0.05$	$\Delta\sigma = 0.1$	$\Delta\sigma = 0.15$	$\Delta\sigma = 0.2$
	0.02	0.05	0.09	0.14
$\Delta\sigma = 0.2$	$\phi = 10°$	$\phi = 20°$	$\phi = 30°$	$\phi = 40°$
	0.02	0.07	0.14	0.18

Table 6.1.: Ratio of in-cusp oriented grain faces $\frac{f_{1/2C}}{f_C}$ that reaches the half-depth width of the cusp.

by the cusp in the energy functional for a constant cusp depth $\Delta\sigma = 0.2$, see figure 6.9. Here again, the dashed line represents the fraction of a random orientation distribution falling in the corresponding orientation space. The graph is complemented by sections through the structure obtained after coarsening from 515 to 250 grains. The observed fractions are more than twice as high as those expected from a random orientation distribution. Energy cusps with a larger opening angle allow a larger variety of low energy orientations, so that a space filling grain boundary network can be build from up to 90% low energy planes. Here too, an increase of straight boundaries and a broader spectrum of dihedral angles is observed for higher fractions f_C of low-energy orientations. Compared with the cusp depth variation, we see slightly more grain boundaries which are curved towards triple lines. This is also due to the greater variety of low-energy orientations.

For all variations of the energy cusp considering torque contributions, the average grain boundary energy was found to decrease during the simulations; a result of the increasing fraction of triangles affected by the cusp. This is in excellent agreement with experimental and simulation results showing a preservation of low energy boundaries during anisotropic coarsening [113, 116]. The average grain boundary energy correlates with both depth and width of the cusp in the energy landscape. From equation 5.6 which gives the proportionality between average grain boundary energy and cusp depth for constant cusp opening angle $\phi = 30°$ and equation 5.7, which gives a similar correlation for the cusp width for constant depth $\Delta\sigma = 0.2$ we can conclude

$$\overline{\gamma} \approx 1 - C \frac{A_C}{A_{tot}} \Delta\sigma. \tag{6.2}$$

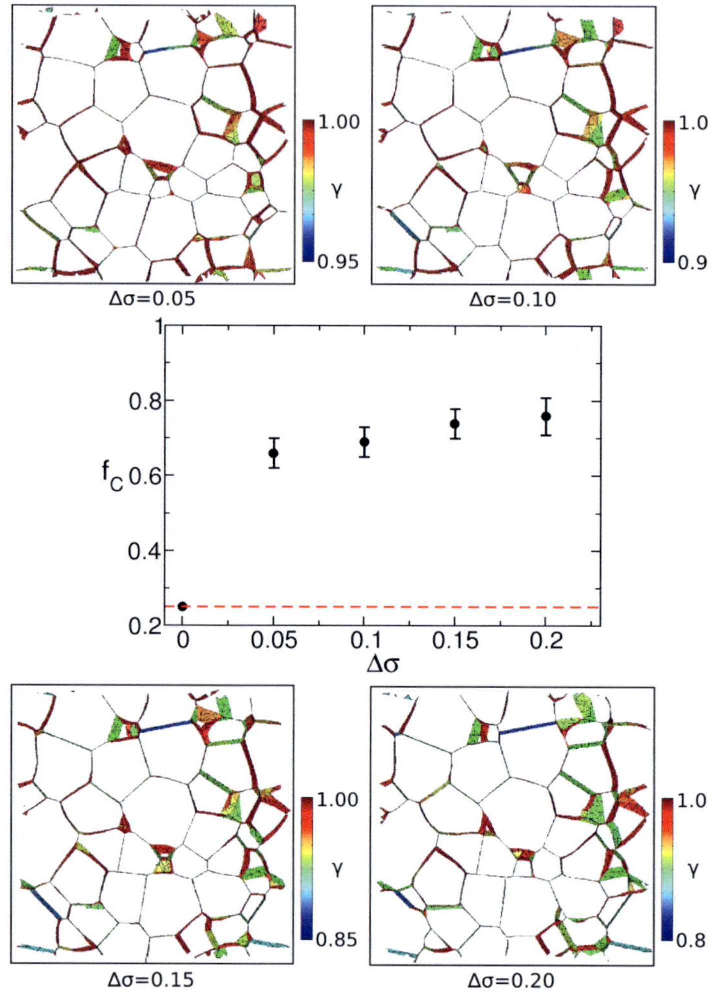

Figure 6.8.: Fraction f_C of discretizational units with in-cusp orientations plotted against cusp depth $\Delta\sigma$ for a constant cusp opening angle $\phi = 30°$. The dashed red line represents the fraction of a random orientation distribution affected by the corresponding area fraction $\frac{A_C}{A_{tot}}$. The inlays show cross sections of the corresponding structures after coarsening to 250 grains.

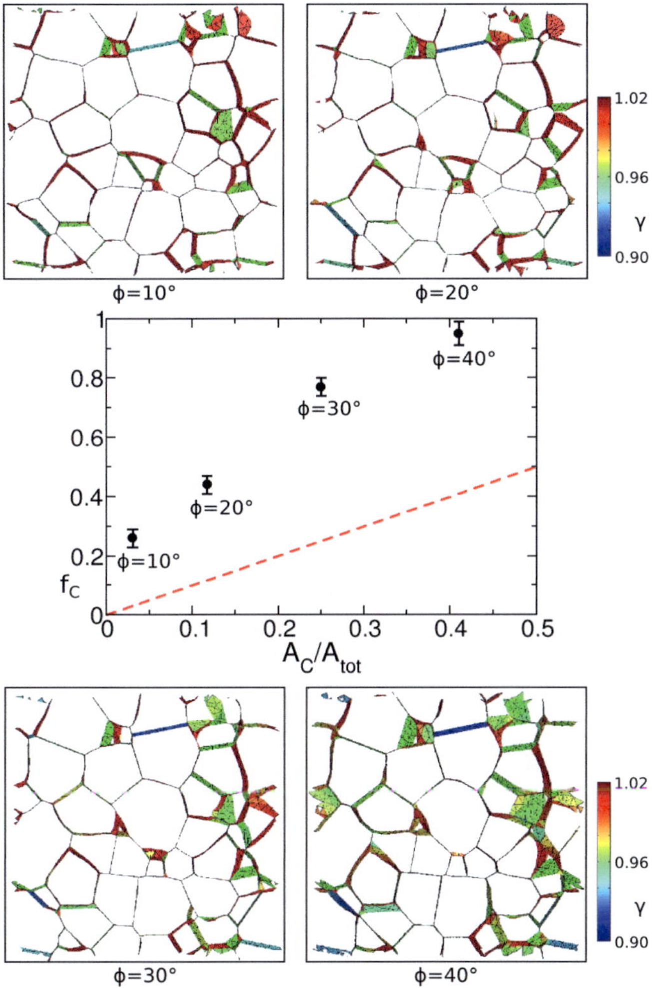

Figure 6.9.: As figure 6.8 for variation of the fraction of the total pole sphere area affected by the cusp in the energy functional $\frac{A_C}{A_{tot}}$.

In the current investigations the constant C is of the order of unity. This is not expected since it implies, that all grain faces with in-cusp orientations receive the full energy advantage, which is not true, since the number of in-cusp orientations that reaches the half-depth width is below 20% for all investigated scenarios, see table 6.1. Therefore we have to assume that the rather large value for C arises from energy gradient effects induced by the variation of cusp depth $\Delta\sigma$ and cusp width ϕ. This assumption is underlined by the fact that the ratio is $\frac{f_{1/C}}{f_C} = 0.02$ for the simulation with $\phi = 30°$, $\Delta\sigma = 0.2$ witout torque contributions. Clearly, this effect must be adressed in more detail by systematic cusp shape variations. However, equation 6.2 does only hold until an upper limit of about 70%-80% of grain faces is affected by the cusp. Afterwards, the average grain boundary energy stagnates. This is rooted in a saturation of the structure with low energy orientations, see also figures 6.8 and 6.9.

Plotting the relative growth rate K as a function of low energy orientations reveals an increasing growth rate that seems to saturate for fractions of low energy triangles greater than 0.7. The absence of data points below $f_C = 0.6$ is rooted in the fact that the cusp depth was investigated for the functional with cusp opening angle $\phi = 30$. Thus the lower limit for the fraction of in-cusp orientations, which would be the random distribution, is given by the area fraction of the pole sphere affected by the cusp $\frac{A_C}{A_{tot}} = 0.25$.

The influence of cusp depth variation and cusp width variation on the growth dynamics is summarized in figure 6.10. Here, the relative growth rate K is parametrized by a cusp variation factor given by the product of area fraction $\frac{A_C}{A_{tot}}$ and cusp depth $\Delta\sigma$. The data points plotted in this diagram reveal a consistent picture, showing an increase in growth rate with rising cusp variation leveling off for $\frac{A_C}{A_{tot}}\Delta\sigma > 0.05$. This is in good agreement with the findings for the average grain boundary reported in equation 6.2. Apparently, the parametrization of the energy anisotropy functional with the proposed cusp variation factor $\left(\frac{A_C}{A_{tot}}\Delta\sigma\right)$ collapsing width and depth of the cusp is appropriate.

The relative growth rate was found to be larger than 1 for all scenarios investigating the effects of energy anisotropy where the average grain boundary energy was smaller than 1, see figure 6.10. This is counter intuitive to the growth law for curvature driven grain growth described by equation 2.14, which suggests a declining growth rate with decreasing average grain boundary energy rooted in a reduced driving force for grain boundary motion. Considering that the average grain boundary energy decreases with increasing cusp variation factor according to equation 6.2, we find the opposite correlation: an increasing growth rate with decreasing average energy. This gives rise to the speculation that a moderate relative energy difference - here of the order of 10-20% -independent of sign leads to an energy gradient which causes the increased growth rate. Under anisotropic conditions the force equilibrium along the interface and especially along triple lines and quadruple points constantly evolves during migration and strongly depends on the interfacial normal. This is different in the isotropic case, where the equilibrium conditions do not evolve and the system stays closer to equilibrium during growth. The net effect of the anisotropic grain boundary energy landscape is therefore an on average increased driving force for grain growth. Furthermore, the grain boundary character distribution is sensitive to grain boundary energy gradients as reported in [15] and observed from our simulations and DCT reconstructions. Together with the increased amount of low energy planes reported both in experiments and simulations, this suggests that also the time evolution of three dimensional grain boundary networks is sensitive to a gradient in grain boundary energy.

In simulation scenarios investigating pure mobility anisotropy no more than the fraction $f_P = 41\%$ of grain boundary orientations expected to be affected by the mobility peak of the corresponding width was observed. Throughout the simulation this value decreased by about 10%, see figure 5.41. This behavior is expected since the high mobility might also enable grains to shrink faster thereby deleting their highly mobile boundaries. Other than in the energy anisotropy scenario, there are no torque contributions rotating new triangles in high mobility orientations. The correlation between average grain boundary mobility $\overline{M}$ and maximum mobility m_{max} described in equation 5.8 can be approximated as a linear mixing rule

$$\overline{M} = (1 - f_P)m_{iso} + f_P M. \tag{6.3}$$

The grain boundary energy for each triangle is built as average of the orientation dependent surface mobilities with respect to both adjacent grains $M = \frac{(m_l + m_r)}{2}$. Since the mobility cusp if of such a form, that either the maximum mobility m_{max} or $m_{iso} = 1$ is allocated and with

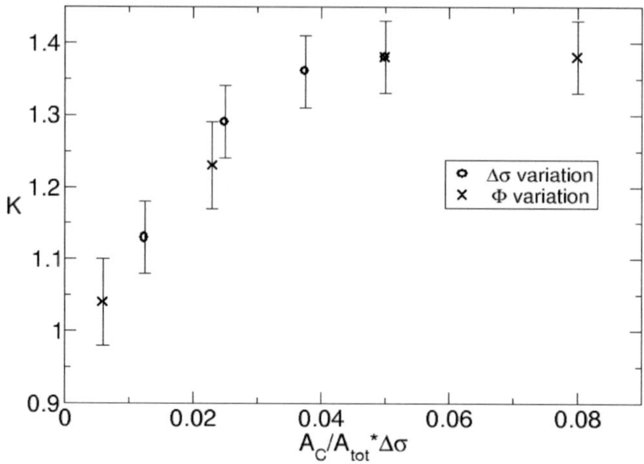

Figure 6.10.: Relative growth rate K plotted against cusp variation factor.

the addition, that for the current investigations only in rare occasions the triangle is oriented such that it represents a high mobility orientation with respect to both adjacent grains, the average mobility can be approximated as

$$\overline{M} = (1 - f_P) + f_P \frac{(m_{max} + 1)}{2}. \tag{6.4}$$

A quadratic correlation between maximum mobility m_{max} and relative growth rate K was observed, see figure 5.43. Together with findings of a simulation scenario varying the fraction f_{GF} of highly mobile grain faces, which are discussed in detail in the context of abnormal grain growth in section 6.3, the impact of a fraction f_{GF} of grain faces provided with a mobility $M = \frac{(m_{max}+1)}{2}$ on the growth dynamics of the system can be described as

$$K = 1 + 0.23 f_{GF} M. \tag{6.5}$$

Simulations combining energy and mobility anisotropy show that providing those grain boundaries affected by the energy cusp with $\Delta\sigma = 0.2$ and $\phi = 30°$ with elevated relative mobility increases the growth rate with respect to the pure energy anisotropy scenario. The increased growth rate is approximately the product of the growth rates for pure energy and pure mobility scenarios. Providing the low energy orientations with lowered relative mobilities on the other hand reduces the growth rate compared to the pure energy anisotropy scenario, see figure 5.46.

In the context of sintering the evolution of pore shapes and sizes during the heating period is a very interesting aspect. During annealing performed in between DCT scans, a significant decrease in volume fraction of porosity (from 2.6% to 1.2%) was observed. This is despite the fact that more pores were detected in the annealed state due to the exploitation of phase contrast tomography data and can thus be regarded as a strong indication for an ongoing densification. Since the material was already dense before heating and we do not observe interconnected pore channels, the densification must be induced by grain boundary diffusion. Unfortunately, a more detailed investigation of pore shape evolution during annealing is ruled out by the comparatively low resolution in the initial state.

6.3. Nucleation of Abnormal Growth

Grain growth simulations investigating the growth behavior of candidate grains as a function of relative grain size and relative grain boundary properties showed that a combined effect of grain boundary energy and mobility advantages is needed to detach an otherwise average grain from a uniform grain size distribution. The results shown in figure 5.53 show that combined energy and mobility advantages can trigger abnormal grain growth in otherwise isotropic grain structures. Compared to a well-accepted mean field approach [36], in our investigations the abnormally growing grains are observed at higher energies and larger relative grain sizes. This gives rise to the assumption that the grain boundary energy has a significant impact on the effective mobility in anisotropic materials. Furthermore, the transition from the normal to abnormal regime for a given energy and mobility ratio is smeared out for the normalized size dependence $\frac{r}{\bar{r}}$. This shows that the three dimensional grain topology also plays an important role for the transition.

The influence of grain boundary mobility anisotropy on the nucleation of abnormal growth was also investigated in simulation scenarios varying the fraction of highly mobile grain faces in otherwise homogeneous

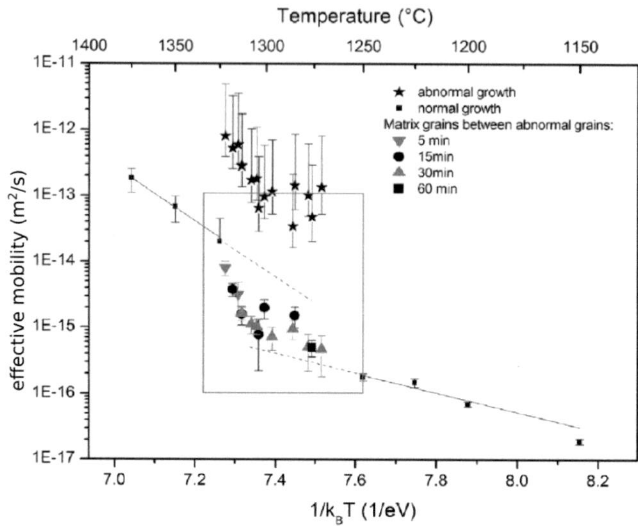

Figure 6.11.: Effective mobility k of the grain boundaries in a polycrystalline $BaTiO_3$ microstructure as a function of the thermal energy [117]. Small markers indicate normal growth behavior, large markers indicate matrix grains between abnormally growing grains as a function of annealing time. Stars denote abnormal grains in the transition region.

grain boundary ensembles. In a simplified system consisting of two grain boundary populations with different mobility, the effective mobility is clearly dependent on the fraction f_{GF} of highly mobile grain faces. For f_{GF} less than 50% a rather moderate increase of the effective mobility and for values larger than 50% a steep increase is observed, see figure 5.44. This simulation result can be related to the transition to a high mobility growth regime and the occurrence of abnormal grain growth observed in $BaTiO_3$ [117]. Figure 6.11 shows the effective mobility k of grain boundaries in a polycrystalline $BaTiO_3$ microstructure as a function of the annealing temperature. Abnormal grain growth occurs in the transition region between the two growth regimes indicated by solid lines. Focusing on the temperature range of the transition zone (between T=1250°C and 1320°C), the growth rate increases at first very little and at higher temperatures a steep increase is observed. In the light of the simulation results presented in section 5.2.1, this transient behavior does not exclude the presence of highly mobile grain boundaries in the intermediate part of the transition zone but strongly suggests that a critical fraction of more than 50% of the grain boundaries have to be in the highly mobile state in order to lead to an overall increase of the effective grain boundary mobility. In the case of an abnormal grain terminated by special orientations this condition is already fulfilled for the faces being part of the abnormal grains boundary only, explaining its accelerated growth.

In order to generate such a configuration the following nucleation mechanism is possible: If there is a preferred low energy orientation close to the orientation of a grain face that undergoes a transition to the high mobility state, the face will tend to turn into the energy minimum due to torque contributions. From that point on, the low boundary energy helps to sustain the grain boundary state during growth. Accordingly a grain that shows no initial size advantage (a grain that lies within the unimodal grain size distribution) needs planes that underwent a transition to a high mobility and have an energy advantage in order to grow abnormally independent of its changing neighboring grains. Grain growth simulations where the fractions of grain faces with pure mobility advantage was varied show a correlation between the fraction of highly mobile faces and growth rate. This correlation happens to be of a nonlinear type revealing a percolation barrier around 50%. Thus a high number of high mobility faces distributed around one growing grain is needed in order to dominate its growth and make it possible for the grain to break free from the log normal distribution observed otherwise.

7. Conclusion

This work focuses on the correlation between interface property anisotropy, microstructure evolution and morphology in polycrystalline perovskites following an integrated experimental and numerical approach. The experiments and simulations presented in the preceding chapters were designed to explore two major questions:

> What is the connection between anisotropic interface properties and the growth anomaly observed in strontium titanate?

> What are the conditions for the nucleation of abnormal grain growth in perovskites?

The applied DCT technique has proven to be an appropriate tool to non-destructively characterize 3D microstructures in ceramics with the added benefit of high resolution grain orientations. The applied instruments and reconstruction techniques yield very good results in terms of image quality, spatial resolution and high accuracy orientation resolution. The goal of non-destructive 3D imaging of microstructure evolution in ceramic materials can be realized with this technique. Even more so, due to the absence of orientation gradients and dislocations undeformed ceramic materials are easier to reconstruct than other polycrystals. Accordingly, grain growth in ceramic materials marks the ideal application area for DCT at its present stage. Despite the possible further improvements in the DCT reconstruction (see Outlook) the currently obtained accuracy in the determination of grain shape and size of about 1.5 µm is accurate enough to directly compare grain growth simulations and DCT observations in $SrTiO_3$ for reasonably large grain growth regimes, provided the initial grain size is large enough. However, the accuracy may not yet be sufficient to identify general anisotropies in grain boundary properties except for very strong cases of faceting.

The modeling approach is confirmed by a validation against analytical predictions for the growth rate. The good agreement in topological quantities obtained experimentally and from grain boundary networks resulting from anisotropic scenarios show the models ability to represent systems with anisotropic grain boundary properties.

Rising fractions of (100) orientations were observed in DCT experiments. At the same time, rising fractions of low energy grain faces were found in vertex dynamics simulations approximating the orientation dependent grain boundary energy as average of the normal dependent surface energies with respect to the crystallography of the two adjacent grains. Together this clearly demonstrates that an energetically favorable orientation with respect to one crystallite is sufficient in strontium titanate to affect the local interface orientation severely. Grain growth simulations applying an orientation dependent grain boundary energy functional based on the normal dependent surface energy functional presented in [82] were found to mimic the growth behavior of $SrTiO_3$ better than simulations including grain boundary mobility anisotropy. From the obtained relationship between average grain boundary energy and cusp variation factor expressed in equation 6.2 we suspect part of the effect we see in the simulations to be attributed to the energy gradient resulting from the sharp cusp in the energy functional. Thus, the cusp shape must be further investigated in order to discriminate, whether an energy advantage of sufficient size or the sharpness of the energy cusp is the dominant effect.

The grain boundary energy anisotropy was identified as the dominant anisotropy in systematic computer simulations. Both topology, as well as growth kinetics are strongly modified by comparatively small changes in relative grain boundary energy. A correlation between fraction of low-energy orientations and relative growth rate was observed. Variation of both the width and the depth of the cusp in the grain boundary energy functional were found to alter the fraction of low-energy orientations. The saturation limit for low-energy orientations was found to be between 70-80% dependent on the cusp shape. Above this limit the growth rate did not increase any farther.

In the context of investigations on the nucleation of abnormal grain growth, a non-linear mixing rule for the effective mobility has been derived from three dimensional grain growth simulations. The mobility measured is below the weighted average of the mobilities present in the microstructure. A critical fraction of roughly 50% of high mobility grain boundaries is needed to change growth dynamics and show significant impact on normal growth, with abnormal grains nucleating with a probability closely linked to the total number of grain boundaries in the high mobility state. A strong coupling of anisotropic grain boundary energy and mobility is needed in order to detach a grain from the log normal grain size distribution. Based on these observations a percolation mechanism for the nucleation of abnormal grain growth was proposed.

8. Outlook

Since time resolved 3D microstructure characterization with complementary crystallographic information became only recently available for polycrystalline ceramic materials in the form of synchrotron based X-ray diffraction contrast tomography experiments, the results presented in this study can only be regarded as a feasibility study and the starting point for a more thorough characterization of microstructure evolution during sintering.

Multiple time step annealing experiments using DCT will not only help to understand the densification during sintering but also give access to a more continuous time evolution of an evolving grain boundary network. Experiments including at least three states will allow to use the first annealed state as confirmation for the correct application of the reconstruction technique and grant comparability of the reconstructions of further annealing states. Recording more time steps would also facilitate comparison with simulations started from the experimentally observed microstructure.

A direct remodeling of synchrotron experiments will allow to study orientation dependent interface properties, especially grain boundary mobility, in greater detail. Systematic parameter variations in grain growth simulations started from 3D microstructure reconstructions will eventually allow to iteratively adapt the grain boundary properties, such that the annealed state is best approximated in the model. This would be a great step towards the modeling based discrimination between orientation and temperature dependent grain boundary energy and mobility. The first prerequisite for this kind of investigations, a conversion of the experimentally obtained structure into an input mesh for the grain growth model is already accomplished, see figure 8.1. In a next step, the influence of the cusp shape on the growth dynamics should be investigated thoroughly. Once isotropic and anisotropic simulations started from these experimentally obtained structures are available, the development of a 3D metric for

Figure 8.1.: DCT reconstruction in the post annealing state transferred to the vertex dynamics model.

comparison of structures evolved in experiment and simulation will be of crucial importance.

The combination of a specific specimen geometry of single crystals bonded to a polycrystalline matrix and DCT will permit observations of the grain boundary mobility with respect to the crystallographic orientation of both sides. Preliminary investigations using SEM imaging showed that the interface mobility is not only temperature but also inclination dependent [2]. These differences in local growth rate are caused by the anisotropy of the mobility and energy with regard to the orientation of the matrix grain, which is not experimentally accessible by conventional metallography. In order to back out the effect of the local driving force, governed by the matrix grain boundaries, the full three dimensional information is needed. Giving access to a statistically relevant set of migration lengths under full knowledge of the crystallography of the participating grains, these experiments will allow generating a database of the interface orientation dependent relative grain boundary mobility.

Prior to future DCT investigations of microstructure in polycrystalline ceramics improvements of the technique could be achieved: The spatial resolution at the grain boundaries can be improved by advanced optical techniques and optimized data processing. This will also lead to improved accuracy with respect to curvature and porosity. Considering the shape of the reconstructed porosities, it is noteworthy, that spherical small pores seem to lie preferably within grains, whilst larger, eccentric shaped pores seem to appear at triple lines. This information could be exploited in an improved version of the dilation algorithm e.g. in creating a probability map for grain and pore areas. High resolution porosity information could consolidate the accuracy of this approach. Better resolved 3D images (0.3 µm pixel size $\sim$ 1µm full width at half maximum of detector point spread function) can be obtained using a different combination of (microscope) optics in the high resolution detector system employed for (parallel beam) PCT measurements. X-ray microscopy techniques (e.g. zoom tomography [118]) can provide higher spatial resolution. A 2048^2 zoom tomography reconstruction with a voxel size of 150 nm would still allow to analyse the 300 µm diameter specimen analysed in the current study. Replacing the up to date uniform dilation step with a more elaborate algorithm based on forward modeling of the diffraction process could represent a starting point for improved microstructure reconstructions with respect to the spatial resolution at the grain boundaries. For each of the unassigned voxels, the goodness of fit of simulated versus measured diffraction patterns can be evaluated while assigning any of the possible (adjacent) grain orientations.

The accuracy of the EBSD validation might be further improved if the material could be in-situ ablated in defined thickness using a machine grinding technique. The recently developed TriBeam technique [69] provides plane-parallel in-situ laser ablation which is significantly faster than the EBSD scan itself and thus allows fast acquisition of large (several hundred slices) 3D post-mortem datasets. The validation against such data is necessary and will help to further improve the reconstruction technique

A. Topological Transformation

The basic elements of the model implementation is described in section 4.1.2. This chapter deals with the topological transformations applied during coarsening of the model structure using the nomenclature as defined in chapter 4.1.2. All topological transformations applied in the vertex dynamics simulation are based on a combination of the following two basic operations:

1. **Joining Vertices** The joining operation is applied to collapse short subedges and merge the connectivity of its endpoints consistently.

   Two joining operations must be distinguished according to the type of the subedge they affect:

   a) subedges with two type 1 vertices as endpoints

   b) subedges with at least one endpoint of type 2 or higher

   Operation 1(a) triggers real topological changes if required, i.e. reconnection of vertices and changes in faces after joining. The steps within this operations are:

   (i) combine the connectivity list of the two endnodes A and B of the subedge and consign this information to the surviving node A.

   (ii) check the subedge or edge connections of the surviving node:

   > remove double connected subedges and collapse the attached triangles to a single subedge.

   > remove one edge of faces shared by the double connected edge.

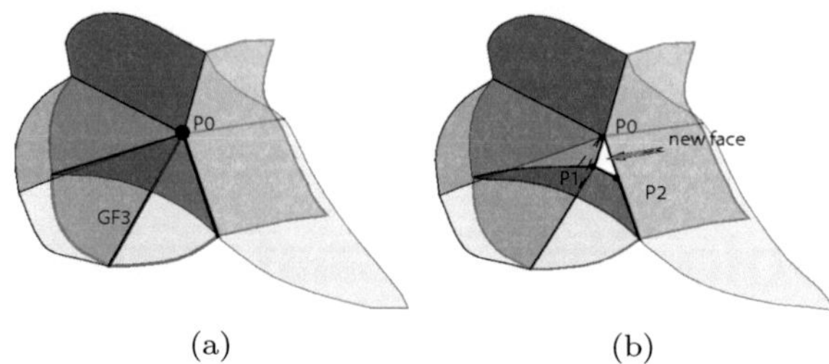

(a)   (b)

Figure A.1.: Detachment process of a grain of type GF3: (a) initial configuration with two edges sharing three faces and one edge sharing four faces; (b) configuration after detachment: a new edge is introduced between the old real vertex $P0$ and the new real vertex $P1$. On each of the outer faces (not shared by GF3) of the detached triple lines now connected to $P1$, the new edge $(P0, P1)$ is added and the face triangulation is adapted. This operation is only performed if a net depinning force of grain GF3 on the vertex $P0$ occurs.

> check all faces sharing node A: remove faces with two or less edges.

(iii) if vertex A is connected to more than three edges check for grain detachment (the second basic topological transformation, detailed description below)

During joining operation 1(b) only steps (i) and (ii) are performed, so that the faces shared by the subedge are rediscretized without affecting the physically relevant topological quantities of the grain.

2. **Detaching a grain from a real vertex**

The detachment procedure is applied to deconnect grains which share two or three faces with a type 1 vertex $P0$. Subsequently, these grains are denoted $GFn(i)$ according to their grain number i and the number of faces n they share with $P0$, see figure 2.

The routine handles edges with three or more shared faces according to the following scheme:

    a) detach all edges and subedges of the selected grain that share the vertex $P0$ and attach them to the new vertex $P1$, which is located at a defined distance $\Gamma = n \cdot l_{crit}$ in direction of the net depinning force.

    b) generate a new edge between vertex $P0$ and $P1$. For the subedges along an edge with endnode $P0$ two cases have to be distinguished:

        i. three faces attached to an edge: insert a new triangle on the outer grain boundary not shared by the detached grain.

        ii. more than three faces attached to an edge: create a new face between grain 1 and 2. Divide the edge into a triple line shared by the detached grain and an edge with one shared face less than the initial edge. The new face is placed between the splitted edge. During this splitting a new type1 vertex $P2$ is introduced at the end of the splitted subedge and new edges are introduced between the vertices $P0, P2$ and $P1, P2$. The new face is spanned by the edges between the three vertices $P0, P1, P2$, see figure 2.

    c) check newly generated edge between $P0$ and $P1$ for consistency after detachment of a grain which shares two faces with $P1$ (edge connectivity test).

Figure 2 illustrates the detachment process for a grain of type $GF3$ sharing one edge with four faces applying step (ii) of the detachment procedure.

The transformation events are triggered based on user defined critical values for basic geometric elements. The following testing scheme checking for the necessity of topological events is applied to the whole data structure at every global time step:

1. check subedges: if the length of a subedge is shorter than $\Delta = f \cdot l_{crit}, f = 0.05, \ldots, 0.1$ and is shrinking or excessively small (shorter than 0.2Δ) apply joining vertices operation to this subedge.

2. check vertices: if vertex has more than three edges search for grain to be detached (the grain with the strongest depinning force).

3. check faces: remove faces with less than three sides if no internal structure/triangulation is present.

4. check grains: remove grains with less than two faces.

In all tests of subedges, vertices and faces, the geometrical objects that meet the critical condition are stored in a temporary list. For the subedges, this list is ordered with increasing subedge length and the application of the topological transformation is done starting with the shortest subedge assuring that transformations will be performed according to their priority.

A second scheme, applied after the motion of the vertices, tests the connectivity of a vertex as follows:

1. check for short subedges attached to the vertex and join vertices if needed.

2. check for deformed triangles shared by the vertex and perform rediscretization if required.

Real vertices connected to more than three edges are tested whether they represent a stable configuration in the grain boundary network. This test is needed as no artificial assumption on the topology of the smallest grain, and on the number of faces shared by one edge, is made in the algorithm. A grain will disappear by loosing faces ending up with a flat grain consisting of two faces. Real vertices sharing three edges occur naturally after a joining operation on real vertices, which may generate edges shared by four faces.

The stability procedure for a real vertex $P0$ consists of the following steps:

1. check the connectivity list of the vertex $P0$: edges sharing two faces only are unphysical as they separate the same grains. Merge the two faces and remove the edge.

2. analyse the grains sharing the vertex $P0$: classify grains according to the number of faces they share with the vertex.

3. test detachment of grains for the general case: start with the class of grains with the lowest number of shared faces. These are the grains with the smallest inner opening angle at the vertex and therefore a presumably large driving force for detachment:

   a) calculate the forces on the real vertex $P0$: a detachment force $\mathbf{f}_{det}^{G}$ resulting from the boundaries of the grain to be detached from vertex $P0$ and the opposing forces $\mathbf{f}_{pin}^{G}$, imposed by the new grain boundaries that need to be created during the detachment process. A depinning is possible, if the condition $(\mathbf{f}_{det}^{G} + \mathbf{f}_{pin}^{G}) \times \mathbf{t}_{in}^{G} > 0$ is fulfilled. Here, $\mathbf{t}_{in}^{G}$ is a vector pointing from the real vertex towards the center of mass of the grain for which detachment is tested.

   b) if more than one grain per class exists, test all possible detachment processes, repeat step (a) for all grains of the class and seclect the one with the highest driving force and best alignment to the direction $\mathbf{t}_{in}^{G}$.

   c) detach selected grain from real vertex (second topological transformation).

   d) check edge between original vertex $P0$ and new vertex $P1$ for connectivity.

A simplification in the detachment routine is done for one of the most frequent configurations consisting of a vertex $P0$ sharing nine different faces and five different grains in total: two grains with a vertex shared by three faces of the grain ($GF3$ type grains), and three grains, with a vertex shared by four faces of the grain ($GF4$ type grains), see figure A.2. This configuration can occur after the collapse of a triple line. If the above described detachment process procedure fails to detach one of the grains of type $GF3$ as the force criterion is not fulfilled for either grain, a new triangular face between the two $GF3$ grains sharing three faces with the vertex $P0$, is generated. This transformation is often referred to as flip. It transfers a shrinking edge to configurations with a triangle between the two $GF3$ grains. The inverse transformation starting from a shrinking triangular face to the generation of a new edge is handled automatically by subsequent application of the node joining transformation, leading temporarily to edges shared by four faces, and finally to a vertex connected

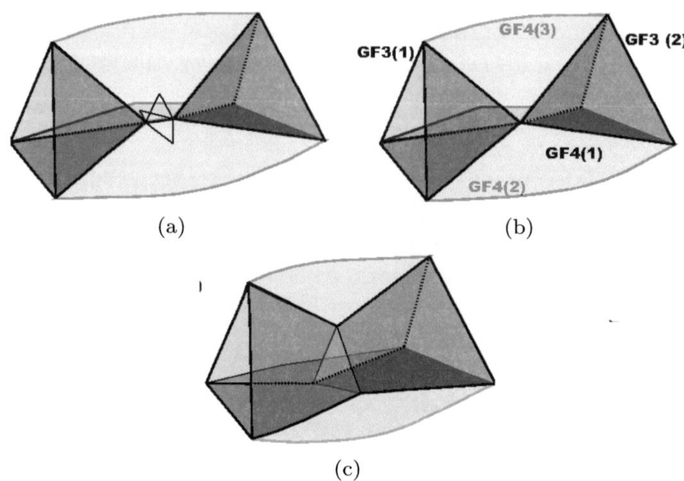

Figure A.2.: Topological transformation: (a) short subedge between two real vertices; (b) intermediate configuration which is decomposed based on force conditions to either the configuration shown in (a) or (c); (c) configuration with a triangular face separating the grains GF3(1) and GF3(2).

to six edges. This vertex is then decomposed using the vertex splitting and grain detachment scheme.

B. Rediscretization

Throughout the simulation a rediscretization procedure is applied to triangles with inner angles smaller or greater than a user defined critical angle ($15°$ and $135°$ in the current configuration) and subedges longer than a critical length ($1.2l_{crit}$), see figure B.1. Depending on the missed quality criteria the following operations are performed :

triangles with inner angles smaller than $15°$ or exceeding $135°$: the shortest subedge are collapsed using the joining routine for vertices, see figure B.1(a)

pair of triangles with inner angles smaller than $15°$ or exceeding $135°$: a swap of subedges of these triangles is performed, see figure B.1(b).

subedges longer than $1.2\,l_{crit}$: a new virtual vertex is introduced along the subedge as shown in figure B.1(c).

The rediscretization scheme does not affect the physical properties of the model: in case of two vertices of different type, the final position is the one of the vertex with the lower type number whereas for vertices of the same type the midpoint position is taken as final position. Furthermore, the operations on subedges of type 1, which are along triple lines, ensure that the triangulation of the other faces remains consistent.

More complex situations consisting of triangles formed by more than one large subedge are also handled: all long subedges of the triangle are divided by introducing virtual vertices and subedges connecting the new vertices are introduced. In the case of three new vertices, three new subedges are introduced, dividing the old triangle into four triangles. For two new vertices one new subedge connects the newly introduced virtual vertices and a further subedge is introduced between a vertex of the old triangle and the new vertex on the opposite subdivided subedge, subdividing the old triangle into three triangles. All the other triangles shared by subdivided subedges are rediscretized accordingly, to ensure a compatible discretization of the concerned faces.

Edge connectivity test

A possible scenario for generating topological redundant edges is detachment of a grain with two faces as the newly created edge has only two faces separating the same grains. In this case the two faces are joined and the edge is removed from the structure. The removal has to be done consistently because the endnodes of the removed edge lose a connection and their topological role changes. Resulting type 1 vertices with two edges are transformed to virtual vertices of type 2 and the shared faces loose one edge.

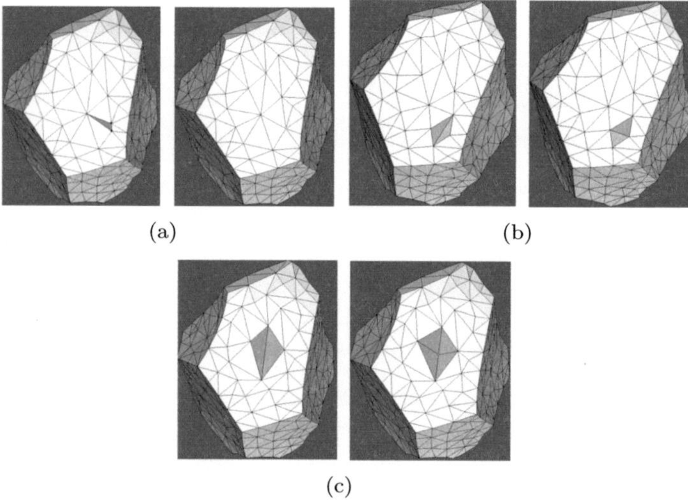

Figure B.1.: Rediscretization: (a) Combine endpoints of a short subedge in ill shaped triangle, (b) switch the common side of a pair of triangles with extreme angles, (c) introduce a discretizational vertex on a long subedge.

Bibliography

[1] M. Bäurer, H. Kungl, and M.J. Hoffmann. Influence of Sr/Ti Stoichiometry on the Densification Behavior of Strontium Titanate. *Journal of the American Ceramic Society*, 92(3):601–606, 2009.

[2] W. Rheinheimer. *Zur Grenzflächenanisotropie von SrTiO3*. Universitätsverlag Karlsruhe, 2013.

[3] D.M. Saylor, B.E. Dasher, T. Sano, and G.S. Rohrer. Distribution of Grain Boundaries in SrTiO3 as a Function of Five Macroscopic Parameters. *Journal of the American Ceramic Society*, 87(4):670–676, 2004.

[4] T. Sano, C.S. Kim, and G.S. Rohrer. Shape Evolution of SrTiO3 Crystals During Coarsening in a Titania-Rich Liquid. *Journal of the American Ceramic Society*, 88(4):993–996, 2005.

[5] S. Shih, K. Dudeck, S.Y. Choi, M. Bäurer, M. Hoffmann, and D. Cockayne. Studies of grain orientations and grain boundaries in polycrystalline SrTiO3.

[6] C. Peng and Y. Chiang. Grain growth in donor-doped SrTiO3. *Journal of Materials Research*, 5(6):1237–1245, 1990.

[7] M. Bäurer, D. Weygand, P. Gumbsch, and M.J. Hoffmann. Grain growth anomaly in strontium titanate. *Scripta Materialia*, 61(6):584 – 587, 2009.

[8] V. Ravikumar, V.P. Dravid, and D. Wolf. Atomic Structure and Properties of the (310) Symmetrical Tilt Grain Boundary (STGB) in SrTiO3. Part I: Atomistic Simulations. *Interface Science*, 8:157–175, 2000.

[9] G. Gottstein and L.S. Shvindlerman. *Grain Boundary Migration in Metals*. CRC Press, Boca Raton, 1999.

[10] W.T. Read and W. Shockley. Dislocation models of crystal grain boundaries. *Scripta Materialia*, 78:275–289, 1950.

[11] A.P. Sutton and R.W. Balluffi. *Interfaces in crystalline materials*. Clarendon Press, Oxford, 1996.

[12] G. Friedel. *Leçons de crystallographie*. Berger-Levrault, Paris, 1926.

[13] D.M. Saylor and G.S. Rohrer. Evaluating Anisotropic Surface Energies using the Capillarity Vector Reconstruction Method. *Interface Science*, 9:35–42, 2001.

[14] G. Wulff. Zur Frage der Geschwindigkeit des Wachstums und der Auflösung der Krystallflächen. *Zeitschrift für Krystallographie und Mineralogie*, 34:449–530, 1901.

[15] G.S. Rohrer. Grain boundary energy anisotropy: a review. *Journal of Materials Science*, 46:5881–5895, 2011.

[16] D.M. Saylor, A. Morawiec, and G.S. Rohrer. The relative free energies of grain boundaries in magnesia as a function of five macroscopic parameters. *Acta Materialia*, 51:3675–3686, 2003.

[17] H. Lueth. *Surfaces and Interfaces of Solids*. Springer, 1993.

[18] A.D. Rollett, G. Gottstein, L. Shvindlerman, and D. Molodov. Grain boundary mobility – a brief review. *Zeitschrift für Metallkunde*, 95(4):226–229, 2004.

[19] M. Upmanyu, D.J. Srolovitz, L.S. Shvindlerman, and G. Gottstein. Misorientation dependence of intrinsic grain boundary mobility: simulation and experiment. *Acta Materialia*, 47(14):3901–3914, 1999.

[20] R.W. Balluffi, S.M. Allen, and W.C. Carter. *Kinetics of Materials*. John Wiley, 2005.

[21] C. Herring. *The Physics of Powder Metallurgy*, pages 143–179. McGraw-Hill, New York, 1951.

[22] C.S. Smith. *Metal Interfaces*, pages 65–108, 1952.

[23] M. Hillert. On the theory of normal and abnormal grain growth. *Acta Metallurgica*, 13(3):227–238, 1965.

[24] D. Weygand, Y. Bréchet, J. Lépinoux, and W. Gust. Three Dimensional Grain Growth: A Vertex Dynamics Simulation. *Philosophical Magazine B*, 79:703–716, 1999.

[25] W.W. Mullins. The statistical self-similarity hypothesis in grain growth and particle coarsening. *Journal of Applied Physics*, 59(4):1341–1349, 1986.

[26] N.P. Louat. On the theory of normal grain growth. *Acta Metallurgica*, 22:721–724, 1974.

[27] P. Feltham. Grain growth in metals. *Acta Metallurgica*, 5:97–105, 1957.

[28] M.F. Vaz and M.A. Fortes. Grain size distribution: The lognormal and the gamma distribution functions. *Scripta Metallurgica*, 22:35–40, 1988.

[29] B. Raeisinia and C.W. Sinclair. A representative grain size for the mechanical response of polycrystals. *Materials Science and Engineering A*, 525:78–82, 2009.

[30] T. Luther and C. Könke. Polycrystal models for the analysis of intergranular crack growth in metallic materials. *Engineering Fracture Mechanics*, 76:2332–2343, 2009.

[31] W.W. Mullins. Two-Dimensional Motion of Idealized Grain Boundaries. *Journal of Applied Physics*, 27(8):900–904, 1956.

[32] J. von Neumann. *Metal Interfaces*, 85:108, 1952.

[33] R.D. MacPherson and D.J. Srolovitz. The von Neumann Relation generalized to coarsening of three-dimensional microstructures. *Nature*, 446:1053–1055, 2007.

[34] C.V. Thompson, H.J. Frost, and F. Spaepen. The relative rates of secondary and normal grain growth. *Acta Metallurgica*, 35:887–890, 1987.

[35] A.D. Rollett, D.J. Srolovitz, and M.P. Anderson. Simulation and Theory of Abnormal Grain Growth-Anisotropic Grain-Boundary Energies and Mobilities. *Acta Metallurgica*, 37:1227–1240, 1989.

[36] F.J. Humphreys. A unified theory of recovery, recrystallization and grain growth, based on the stability and growth of cellular microstructures – I. The basic model. *Acta Materialia*, 45:4231–4240, 1997.

[37] P.R. Rios. Abnormal grain growth development from uniform grain size distributions. *Acta Materialia*, 45:1785–1789, 1997.

[38] P.R. Rios. Abnormal grain growth development from uniform grain size distributions due to a mobility advantage. *Scripta Materialia*, 38:1359–1364, 1998.

[39] P.R. Rios and M.E. Glicksman. Topological theory of abnormal grain growth. *Acta Materialia*, 54:5313–5321, 2006.

[40] N. Chen, S. Zaefferer, L. Lahn, K. Günther, and D. Raabe. Effects of topology on abnormal grain growth in silicon steel. *Acta Materialia*, 51:1755–1765, 2003.

[41] M.E. Glicksman. Analysis of 3-D network structures. *Philosophical Magazine*, 85:3–31, 2005.

[42] P.R. Rios and M.E. Glicksman. Self-similar evolution of network structures. *Acta Materialia*, 54:1041–1051, 2006.

[43] M.P. Anderson, G.S. Grest, P.S. Sahni, and D.J. Srolovitz. Computer simulation of grain growth–I. Kinetics. *Acta Metallurgica*, 32(5):783–791, 1984.

[44] M.W. Nordbakke, N. Ryum, and O. Hunderi. Invariant distributions and stationary correlation functions of simulated grain growth processes. *Acta Materialia*, 50:3661–3670, 2002.

[45] C. Wang and G. Liu. On the stability of grain structure with initial Weibull grain size distribution. *Materials Letters*, 57(28):4424–4428, 2003.

[46] O.M. Ivasishin, S.V. Shevchenko, N.L. Vasiliev, and S.L. Semiatin. 3D Monte-Carlo Simulation of texture-controlled grain growth. *Acta Materialia*, 51:1019–1034, 2003.

[47] D. Zoellner. *Monte Carlo Potts Model Simulation and Statistical Mean-Field Theory of Normal Grain Growth*. Otto-von-Guericke-Universität, Magdeburg, 2006.

[48] N. Metropolis and S. Ulam. The Monte Carlo Method. *Journal of the American Statistical Association*, 44:335–341, 1949.

[49] J.B. Collins and H. Levine. Diffuse interface model of diffusion-limited crystal growth. *Physical Review B*, 31:6119–6122, 1985.

[50] C. E. Krill III and L. Q. Chen. Computer simulation of 3-D grain growth using a phase-field model. *Acta Materialia*, 50(12):3059–3075, 2002.

[51] Y. Suwa, Y. Saito, and H. Onodera. Three-dimensional phase field simulation of the effect of anisotropy in grain-boundary mobility on growth kinetics and morphology of grain structure. *Computational Materials Science*, 40(1):40 – 50, 2007.

[52] N. Moelans, B. Blanpain, and P. Wollants. Quantitative Phase-Field Approach for Simulating Grain Growth in Anisotropic Systems with Arbitrary Inclination and Misorientation Dependence. *Physical Review Letters*, 101:025502, 2008.

[53] K. Kawasaki, T. Nagai, and K. Nakashima. Vertex models for two-dimensional grain growth. *Philosophical Magazine B*, 60:399–421, 1989.

[54] M. Syha and D. Weygand. A generalized vertex dynamics model for grain growth in three dimensions. *Modelling and Simulation in Materials Science and Engineering*, 18(1):015010, 2010.

[55] F. Wakai, N. Enomoto, and H. Ogawa. Three-dimensional microstructural evolution in ideal grain growth–general statistics. *Acta Materialia*, 48(6):1297–1311, 2000.

[56] D. Weygand, Y. Bréchot, and J. Lépinoux. Reduced Mobility of Triple Nodes and Lines on Grain Growth in Two and Three Dimensions. *Interface Science*, 7:285–295, 1999.

[57] L.A. Barrales Mora, V. Mohles, L.S. Shvindlerman, and G. Gottstein. Effect of a finite quadruple junction mobility on grain microstructure evolution: Theory and simulation. *Acta Materialia*, 56(5):1151–1164, 2008.

[58] M. Syha, M. Bäurer, M.J. Hoffmann, E.M. Lauridsen, W. Ludwig, D.Weygand, and P. Gumbsch. Comparing Grain Growth Experiments and Simulations in 3D. *Proceedings of the 31st Risø International Symposium on Materials Science*, 1:138–140, 2010.

[59] M. Syha and D. Weygand. Conditions for the Occurence of Abnormal Grain Growth studied by a 3D Vertex Dynamics Model. *Materials Science Forum*, 715:563–567, 2012.

[60] K. Ahmed, C.A. Yablinsky, A. Schulte, T. Allen, and A. El-Azab. Phase field modeling of the effect of porosity on grain growth kinetics in polycrystalline ceramics. *Modelling and Simulation in Materials Science and Engineering*, 21(6):065005, 2013.

[61] I. Steinbach, X. Song, and A. Hartmaier. Phase-field model with plastic flow for grain growth in nanocrystalline material. *Philosophical Magazine*, 90:485–499, 2010.

[62] N. Zhou, C. Shen, M.J. Mills, and Y. Wang. Phase field modeling of channel dislocation activity and γ rafting in single crystal Ni–Al. *Acta Materialia*, 55:5369–5381, 2007.

[63] M. Miodownik, E.A. Holm, and G.N. Hassold. Highly parallel computer simulations of particle pinning: Zener vindicated. *Scripta Materialia*, 42:1173–1177, 2000.

[64] O.M. Ivasishin, S.V. Shevchenko, N.L. Vasiliev, and S.L. Semiatin. A 3-D Monte-Carlo (Potts) model for recrystallization and grain growth in polycrystalline materials. *Materials Science and Engineering A*, 433:216–232, 2006.

[65] R.T. DeHoff. Quantitative serial sectioning analysis. *Journal of Microscopy*, 131:259–263, 1983.

[66] J.E. Spowart, H.M. Mullens, and B.T. Puchala. Collecting and Analyzing Microstructures in Three Dimensions: A Fully Automated

Approach. *Journal of the Minerals, Metals and Materials Society*, 55(10):35–37, 2003.

[67] T. Sakamoto, Z. Cheng, M. Takahasi, M. Owari, and Y. Nihei. Development of an ion and electron dual focused beam apparatus for three-dimensional microanalysis. *Japanese Journal of Applied Physics*, 37:2051–2056, 1998.

[68] M.A. Groeber, B.K. Haley, M.D. Uchic, D.M. Dimiduk, and S. Ghosh. 3D reconstruction and characterization of polycrystalline microstructures using a FIB–SEM system. *Materials Characterization*, 57:259–273, 2006.

[69] M.P. Echlin, A. Mottura, C.J. Torbet, and T.M. Pollock. A new TriBeam system for three-dimensional multimodal analysis. *Review of Scientific Instruments*, 83:023701, 2012.

[70] E.M. Lauridsen, S. Schmidt, L. Margulies, H.F. Poulsen, and D. Juul Jensen. Growth kinetics of individual cube grains as studied by the 3D X-ray diffraction microscope. *Recrystallization and Grain Growth, Proceedings of the First Joint International Conference*, 1:589–594, 2001.

[71] H.F. Poulsen. *Three-dimensional X-Ray diffraction microscopy. Mapping polycrystals and their dynamics.* Springer, 2004.

[72] R.M. Suter, D. Hennessy, C. Xiao, and U. Lienert. Forward modeling method for microstructure reconstruction using X-ray diffraction microscopy: Single-crystal verification. *Review of Scientific Instruments*, 77:123905, 2006.

[73] G. Johnson, A. King, M.G. Honnicke, J. Marrow, and W. Ludwig. X-ray diffraction contrast tomography: a novel technique for three-dimensional grain mapping of polycrystals. II. The combined case. *Journal of Applied Crystallography*, 41:310–318, 2008.

[74] W. Ludwig, P. Reischig, A. King, M. Herbig, E.M. Lauridsen, G. Johnson, T.J. Marrow, and J.Y. Buffière. Three-dimensional grain mapping by X-ray diffraction contrast tomography and the use of Friedel pairs in diffraction data analysis. *Review of Scientific Instruments*, 80(3):033905, 2009.

[75] B.C. Larson, W. Yang, G.E. Ice, J.D. Budai, and J.Z. Tischler. Three-dimensional X-ray structural microscopy with submicrometre resolution. *Nature*, 415:887–890, 2002.

[76] J. Baruchel, J.Y. Buffière, P. Cloetens, M. Di Michiel, E. Ferrie, W. Ludwig, E. Maire, and L. Salvo. Advances in synchrotron radiation microtomography. *Scripta Materialia*, 55(1):41–46, 2006.

[77] K.J. Batenburg, J. Sijbers, H.F. Poulsen, and E. Knudsen. DART: a robust algorithm for fast reconstruction of three-dimensional grain maps. *Journal of Applied Crystallography*, 43:1464, 2010.

[78] A.C. Kak and M. Slaney. *Principles of computerized tomographic imaging*. IEEE Press, 1987.

[79] F.J. Humphreys and I. Brough. 1. High resolution electron backscatter diffraction with a field emission gun scanning electron microscope. *Journal of Microscopy*, 195:6–9, 1999.

[80] T.C. Isabell and V.P. Dravid. Resolution and sensitivity of electron backscattered diffraction in a cold field emission gun SEM. *Ultramicroscopy*, 67:59–68, 1997.

[81] S. Zaefferer. On the formation mechanisms, spatial resolution and intensity of backscatter Kikuchi patterns. *Ultramicroscopy*, 107:254–266, 2007.

[82] T. Sano, D.M. Saylor, and G.S. Rohrer. Surface Energy Anisotropy of $SrTiO_3$ at 1400 °C in Air. *Journal of the American Ceramic Society*, 86(11):1933–1939, 2003.

[83] R.O. Bell and G. Rupprecht. Elastic Constants of Strontium Titanate. *Phyisical Review*, 129:90–94, 1963.

[84] M. Bäurer. *Kornwachstum in Strontiumtitanat*. Universitätsverlag Karlsruhe, 2009.

[85] M. Bäurer, S.J. Shih, C. Bishop, M.P. Harmer, D. Cockayne, and M.J. Hoffmann. Abnormal grain growth in undoped strontium and barium titanate. *Acta Materialia*, 58:290–300, 2010.

[86] S.Y. Chung and S.L. Kang. Effect of Dislocations on Grain Growth in Strontium Titanate. *Journal of the American Ceramic Society*, 83:2828–2832, 2000.

[87] S.Y. Chung and S.L. Kang. Intergranular amorphous films and dislocations-promoted grain growth in $SrTiO_3$. *Acta Materialia*, 51:2345–2354, 2003.

[88] C. Herring. Some theorems on the free energies of crystal surfaces. *Physical Review*, 82:87–93, 1951.

[89] Z. Wu and J.M. Sullivan. Multiple material marching cubes algorithm. *International Journal for Numerical Methods in Engineering*, 58:189–207, 2003.

[90] J.K. Mackenzie. Second Paper on Statistics Associated with the Random Disorientation of Cubes. *Biometrika*, 45:229–240, 1958.

[91] M.A. Groeber, S. Ghosh, M.D. Uchic, and D.M. Dimiduk. A framework for automated analysis and simulation of 3D polycrystalline microstructures. Part1: Statistical characterization. *Acta Materialia*, 56:1257–1273, 2008.

[92] P.E. Danielsson. Euclidean Distance Mapping. *Computer Graphics and Image Processing*, 14:227–248, 1980.

[93] W.W. Mullins. Estimation of the geometrical rate constant in idealized three dimensional grain growth. *Acta Metallurgica*, 37(11):2979–2984, 1989.

[94] L.A. Barrales Mora, G. Gottstein, and L.S. Shvindlerman. Three-dimensional grain growth: Analytical approaches and computer simulations. *Acta Materialia*, 56(20):5915–5926, 2008.

[95] E.E. Underwood. The Mathematical Foundations of Quantitative Stereology. *Stereology and Quantitative Metallography, ASTM STP*, 504:3–38, 1972.

[96] S.A. Saltykov. *Stereometric metallography*. State Publishing House for Metals and Sciences, Moscow, 1958.

[97] P. Cloetens, M. Pateyron-Salomé, J. Y. Buffière, G. Peix, J. Baruchel, F. Peyrin, and M. Schlenker. Observation of microstructure and damage in materials by phase sensitive radiography and tomography. *Journal of Applied Physics*, 81:5878, 1997.

[98] H.S.M. Coxeter. *Regular Polytopes*. Macmillan Company, 1987.

[99] C.S. Smith. Some elementary principles of polycrystalline microstructure. *Metallurgical Reviews*, 9:1–48, 1964.

[100] R.E. Williams. Space-Filling Polyhedron: Its Relation to Aggregates of Soap Bubbles, Plant Cells, and Metal Crystallites. *Science*, 161(3838):276–277, 1968.

[101] R.T. DeHoff and G.Q. Liu. On the relation between grain size and grain topology. *Metallurgical Transactions A*, 16:2007–2011, 1985.

[102] G. Liu, H. Yu, and Y. Qin. Three-dimensional grain topology-size relationships in a real metallic polycrystal compared with theoretical models. *Materials Science and Engineering A*, 326:276–281, 2002.

[103] J. Padilla and D. Vanderbilt. Ab-initio study of $SrTiO_3$ surfaces. *Surface Science*, 418:64–70, 1998.

[104] A. Pojani, F. Finocchi, and C. Noguera. Polarity on the $SrTiO_3$ (111) and (110) surfaces. *Surface Science*, 442:179–198, 1999.

[105] D.M. Saylor, B.E. Dasher, Y. Pang, H.M. Miller, P. Wynblatt, A.D. Rollett, and G.S. Rohrer. Habits of Grains in Dense Polycrystalline Solids. *Journal of the American Ceramic Society*, 87(4):724–726, 2004.

[106] S. Shih, S. Lozano-Perez, and D. Cockayne. Investigation of grain boudnaries for abnormal grain growth in polycrystalline $SrTiO_3$.

[107] M. Bäurer, H. Störmer, D. Gerthsen, and M.J. Hoffmann. Linking Grain Boundaries and Grain Growth in Ceramics. *Advanced Engineering Materials*, 12:1230–1234, 2010.

[108] E. Ernst, M.L. Mulvihill, O. Kienzle, and M. Rühle. Preferred grain orientation relationships in sintered perovskite ceramics. *Journal of the American Ceramic Society*, 84:1885–1890, 2001.

[109] I. Nettleship, R.J. McAfee, and W.S. Slaughter. Evolution of the Grain Size Distribution during the Sintering of Alumina at 1350°C. *Journal of the American Ceramic Society*, 85:1954–1960, 2002.

[110] Y.H. Heo, S.C. Jeon, J.G. Fisher, S.Y. Choi, K.H. Hur, and S.J. Kang. Effect of step free energy on delayed abnormal grain growth in a liquid-phase sintered $BaTiO_3$ model system. *Journal of the European Ceramic Society*, 31:755–762, 2011.

[111] A. Kazaryan, Y. Wang, S.A. Dregia, and B.R. Patton. Grain growth in anisotropic systems: comparison of effects of energy and mobility. *Acta Materialia*, 50:2491–2502, 2002.

[112] E.A. Holm and C.C. Bataille. The Computer Simulation of Microstructural Evolution. *Journal of the Minerals, Metals and Materials Society*, 53(9):20–23, 2012.

[113] E.A. Holm, G.N. Hassold, and M.A. Miodownik. On misorientation distribution evolution during anisotropic growth. *Acta Materialia*, 49:2981–2991, 2001.

[114] R.M. German. Coarsening in Sintering: Grain Shape Distribution, Grain Size Distribution, and Grain Growth Kinetics in Solid-Pore Systems. *Critical Reviews in Solid State and Material Sciences*, 35(4):263–305, 2010.

[115] A. Tewari, A.M. Gokhale, and R.M. German. Effect of gravity on three-dimensional coordination number distribution in liquid phase sintered microstructures. *Acta Materialia*, 47:3721–3734, 1999.

[116] M. Upmanyu, G.N. Hassold, A. Kazaryan, E.A. Holm, Y. Wang, B. Patton, and D.J. Srolovitz. Boundary Mobility and Energy Anisotropy Effects on Microstructural Evolution During Grain Growth. *Interface Science*, 10:201–216, 2002.

[117] M. Bäurer, M. Syha, and D. Weygand. Combined experimental and numerical study on the effective grain growth dynamics in highly anisotropic systems: Application to Barium Titanate. *Acta Materialia*, 61:5664–5673, 2013.

[118] R. Mokso, P. Cloetens, E. Maire, W. Ludwig, and J.-Y. Buffière. Nanoscale zoom tomography with hard X-rays using Kirkpatrick-Baez optics. *Applied Physics Letters*, 90:144104, 2007.

Schriftenreihe
des Instituts für Angewandte Materialien

ISSN 2192-9963

Die Bände sind unter www.ksp.kit.edu als PDF frei verfügbar
oder als Druckausgabe bestellbar.

Band 1   Prachai Norajitra
Divertor Development for a Future Fusion Power Plant. 2011
ISBN 978-3-86644-738-7

Band 2   Jürgen Prokop
**Entwicklung von Spritzgießsonderverfahren zur Herstellung
von Mikrobauteilen durch galvanische Replikation.** 2011
ISBN 978-3-86644-755-4

Band 3   Theo Fett
**New contributions to R-curves and bridging stresses –
Applications of weight functions.** 2012
ISBN 978-3-86644-836-0

Band 4   Jérôme Acker
**Einfluss des Alkali/Niob-Verhältnisses und der Kupferdotierung
auf das Sinterverhalten, die Strukturbildung und die Mikrostruktur von bleifreier Piezokeramik $(K_{0,5}Na_{0,5})NbO_3$.** 2012
ISBN 978-3-86644-867-4

Band 5   Holger Schwaab
**Nichtlineare Modellierung von Ferroelektrika unter
Berücksichtigung der elektrischen Leitfähigkeit.** 2012
ISBN 978-3-86644-869-8

Band 6   Christian Dethloff
**Modeling of Helium Bubble Nucleation and Growth
in Neutron Irradiated RAFM Steels.** 2012
ISBN 978-3-86644-901-5

Band 7   Jens Reiser
**Duktilisierung von Wolfram. Synthese, Analyse und
Charakterisierung von Wolframlaminaten aus Wolframfolie.** 2012
ISBN 978-3-86644-902-2

Band 8   Andreas Sedlmayr
**Experimental Investigations of Deformation Pathways
in Nanowires.** 2012
ISBN 978-3-86644-905-3

Band 29   Gunnar Picht
           **Einfluss der Korngröße auf ferroelektrische Eigenschaften
           dotierter $Pb(Zr_{1-x}Ti_x)O_3$ Materialien.** 2013
           ISBN 978-3-7315-0106-0

Band 30   Esther Held
           **Eigenspannungsanalyse an Schichtverbunden
           mittels inkrementeller Bohrlochmethode.** 2013
           ISBN 978-3-7315-0127-5

Band 31   Pei He
           **On the structure-property correlation and the evolution
           of Nanofeatures in 12-13.5% Cr oxide dispersion strengthened
           ferritic steels.** 2014
           ISBN 978-3-7315-0141-1

Band 32   Jan Hoffmann
           **Ferritische ODS-Stähle – Herstellung,
           Umformung und Strukturanalyse.** 2014
           ISBN 978-3-7315-0157-2

Band 33   Wiebke Sittel
           **Entwicklung und Optimierung des Diffusionsschweißens
           von ODS Legierungen.** 2014
           ISBN 978-3-7315-0182-4

Band 34   Osama Khalil
           **Isothermes Kurzzeitermüdungsverhalten der hoch-warmfesten
           Aluminium-Knetlegierung 2618A (AlCu2Mg1,5Ni).** 2014
           ISBN 978-3-7315-0208-1

Band 35   Magalie Huttin
           **Phase-field modeling of the influence of mechanical stresses on
           charging and discharging processes in lithium ion batteries.** 2014
           ISBN 978-3-7315-0213-5

Band 36   Christoph Hage
           Grundlegende Aspekte des 2K-Metallpulverspritzgießens. 2014
           ISBN 978-3-7315-0217-3

Band 37   Bartłomiej Albiński
           **Instrumentierte Eindringprüfung bei Hochtemperatur
           für die Charakterisierung bestrahlter Materialien.** 2014
           ISBN 978-3-7315-0221-0

Band 38   Tim Feser
           **Untersuchungen zum Einlaufverhalten binärer alpha-
           Messinglegierungen unter Ölschmierung in Abhängigkeit
           des Zinkgehaltes.** 2014
           ISBN 978-3-7315-0224-1

Band 39   Jörg Ettrich
Fluid Flow and Heat Transfer in Cellular Solids. 2014
ISBN 978-3-7315-0241-8

Band 40   Melanie Syha
Microstructure evolution in strontium titanate Investigated by means of grain growth simulations and x-ray diffraction contrast tomography experiments. 2014
ISBN 978-3-7315-0242-5